PREPARING
FOR THE
GREAT WAVES
OF CHANGE

PREPARING
FOR THE
GREAT WAVES
OF CHANGE

HOW TO LIVE
IN A CHANGING WORLD

AS REVEALED TO
Marshall Vian Summers

PREPARING
FOR THE
GREAT WAVES
OF CHANGE

Copyright © 2025 by The Society for the New Message

All rights reserved. No part of this publication may be reproduced, stored in a retrieval system or transmitted in any form or by any means: electronic, mechanical, photocopying, recording or otherwise without the prior written permission of the publisher.

Edited by Darlene Mitchell
Cover design by Tyyne Andrews
Interior design by Virginia Sender

ISBN: 978-1-942293-33-0 (POD)
ISBN: 978-1-942293-34-7 (Ebook)
NKL POD Version 7.1 ; Sv.7.9
Library of Congress Control Number: 2024920316

Publisher's Cataloging-in-Publication
(Provided by Cassidy Cataloging Services, Inc)

Names:	Summers, Marshall Vian, 1949- author. | Society for the New Message, issuing body.
Title:	Preparing for the great waves of change : how to live in a changing world / as revealed to Marshall Vian Summers.
Description:	Boulder, CO : The Society for the New Message, [2024]
Identifiers:	ISBN: 978-1-942293-33-0 (POD) | 978-1-942293-34-7 (ebook) | LCCN: 2024920316
Subjects:	LCSH: Society for the New Message--Doctrines. | Change--Religious aspects. | Religion and social problems. | Knowledge, Theory of (Religion) | Extraterrestrial beings. | Spiritual life. | Social evolution. | Summers, Marshall Vian, 1949- Religion.
Classification:	LCC: BP605.S58 S869 2024 | DDC: 299/.93--dc23

Preparing for the Great Waves of Change is a book of the New Message and is published by New Knowledge Library, the publishing imprint of The Society for the New Message. The Society is a religious non-profit organization dedicated to teaching and publishing the New Message. The books of New Knowledge Library can be ordered at www.newknowledgelibrary.org, your local bookstore and at many other online retailers.

The New Message is being studied in more than 30 languages in over 90 countries. *Preparing for the Great Waves of Change* is being translated into many languages of our world by dedicated volunteer translators from around the world. These translations will all be available online at www.newmessage.org.

The Society for the New Message
P.O. Box 1724   Boulder, CO 80306-1724
(303) 938-8401   (800) 938-3891
011 303 938 84 01   (International)
newmessage.org   newknowledgelibrary.org
email: society@newmessage.org

*If you can begin to prepare,
you will gain confidence.
If you do nothing, you will lose heart.*

From *The Great Waves of Change*
Chapter 8: The Danger of Isolation

---

*Environment is the most important thing.
When it changes, everything changes.*

Marshall Vian Summers

PREPARING
FOR THE
GREAT WAVES
OF CHANGE

TABLE OF CONTENTS

Introduction i

CHAPTER 1	Navigating the Difficult Times Ahead.................	1
CHAPTER 2	Facing a World in Decline..................................	21
CHAPTER 3	Living in a Time of Uncertainty and Instability....	35
CHAPTER 4	Preparing for the Future	53
CHAPTER 5	Adapting to a Changing World	71
CHAPTER 6	Escaping Fear, Confusion and Hopelessness.....	85
CHAPTER 7	Who Can You Trust?...	103
CHAPTER 8	How Will You Know What to Do?.......................	121
CHAPTER 9	Realizing the Need to Prepare	137
CHAPTER 10	Seeing What Is Coming	153
CHAPTER 11	Signs from the World...	167
CHAPTER 12	Self-Reliance	183
CHAPTER 13	Building Your Ark...	199
CHAPTER 14	Finding Certainty and Strength	215
CHAPTER 15	The Greater Mind of Knowledge........................	231

Important Terms 249
About Marshall Vian Summers 257
The Voice of Revelation 259
About The Society for the New Message 261
Books of the New Message.. 263

Introduction

$\mathcal{A}$s we open the pages of *Preparing for the Great Waves of Change*, we are confronted with several new challenges and concepts for us as individuals and for humanity as a whole about the world we live in and about the universe around us.

As a world, we are now facing two significant challenges. The first is what this book calls the "Great Waves of change," which are a converging set of social, environmental, and economic forces impacting the world. These converging and accelerating forces are the result of centuries of environmental destruction and misuse of the world which pose a major threat to the well-being of people everywhere, even in the most wealthy nations.

The second is Contact with a "Greater Community" of intelligent life in the universe. Visitation by extraterrestrial life from beyond our solar system is now underway around the world and represents a major step in the evolution of the human species and a significant challenge to our human sovereignty. We are no longer alone in the universe or even within our own world.

These two realities can initially be quite shocking. I know firsthand the shock of encountering the reality of what we are facing. Perhaps you too have felt that something has changed in the world even though things may look much as they always have. If so, you are not alone. Many people all around the world are experiencing a vague or sometimes acute sense of anxiety, apprehension, and uneasiness about the future.

When first encountering Marshall Vian Summers's teachings, I was immediately struck by their profound depth and personal resonance. Though I couldn't fully grasp their significance at the time, there was an unmistakable feeling of authenticity that drew me in. As I delved further into his work, I found myself deeply moved by the clarity and insight this book provides on navigating the complexities

PREPARING FOR THE GREAT WAVES OF CHANGE

and challenges of modern human existence, and how to contribute meaningfully to this world.

Preparing for the Great Waves of Change is part of a larger communication received by Marshall Vian Summers from a group of spiritual teachers offering a new awareness and preparation for navigating the challenges of our rapidly changing world. Through his deep connection with these spiritual guides, he has unveiled a new message for humanity, currently comprised of twenty books in publication, with many more to come. This message provides a direct experience of spiritual reality and offers a profound understanding of God that transcends traditional religious boundaries.

What resonates most with me is the practical wisdom and guidance found within these teachings, which address nearly every aspect of human existence. What is offered here is not simply a spiritual philosophy but a comprehensive roadmap for living a fulfilling and purposeful life in today's complex world. For me, it feels like coming home to a truth that has long been dormant waiting to be rediscovered.

Preparing for the Great Waves of Change is a book of teachings to help us discover what the signs of our changing world may be telling us at a deeper level. Feelings of anxiety, apprehension, and uneasiness can be indicators of a spiritual intelligence within us that is trying to get our attention and flag us down as we barrel headlong into a future that will be much more challenging.

Preparing for the Great Waves of Change encourages us to bring forth the gift of our innate Knowledge—not intellectual knowledge but an innate human intelligence attempting to guide us in life. This is our ability to know things directly beyond reason and is a guide for each of us to navigate and offset these forces and prepare wisely for their impacts.

It is about undertaking an inner and outer preparation in our lives. But inner preparation is primary and essential in guiding our outer preparation, for we cannot stockpile enough food or barricade

INTRODUCTION

or arm ourselves for every eventuality. But we can develop our innate intelligence to guide our actions step by step as we face the alarming likelihood of escalating food costs and shortages, economic hardship, rising sea levels, catastrophic weather events, and conflict and war over the world's remaining resources, all while facing competitive visitation from the universe.

Given all that we are facing, imagine what it would mean to have an unshakeable foundation in life to guide us no matter what erupts or breaks down in our outer world. The preparation offered here can give us this foundation and the ability to remain calm, centered, and adaptable in the face of long-term or immediate crises.

Here, *Preparing for the Great Waves of Change* contains a method of inquiry and evaluation that could potentially be life saving. Questions are posed and action steps are presented to assist us in the process. Even though these two challenges have the power to degrade humanity's well-being, freedom, and sovereignty, they also hold the power of redemption and advancement. Over time, I have discovered the strength and guidance offered within these pages, which people everywhere can receive as we navigate this new world and confront its challenging realities.

In this book, we are given much to consider and be with now and over time to unearth the movement of the world and the movement needed in our lives to position ourselves as contributors instead of casualties in this rapidly changing environment. We are given the opportunity to look at the world with fresh eyes—to take the blinders off and see the signs the world itself is giving us about what is coming.

Preparing for the Great Waves of Change states that we are now at the end of the age of indulgence. The times now require human unity, cooperation, and immense courage and ingenuity, not from a few saintly or gifted people but from each of us all around the world from diverse cultures and life experiences.

PREPARING FOR THE GREAT WAVES OF CHANGE

Read and explore the awareness, education, and preparation held within this book for living in and contributing to a Great Waves world. Remember that the change is not out there somewhere. It is right here within us, in our ability to see, to know, and to act according to our deeper spiritual nature and through integrating our thinking mind and our spiritual mind into a higher functioning whole.

The real inspiration and promise for humanity is in the awakening of each person and our resulting actions as individuals and as a whole human family.

Tyyne Andrews
June 2024

CHAPTER 1

Navigating the Difficult Times Ahead

Humanity is entering a time of great change and difficulty, a time when it will be facing the Great Waves of change: environmental degradation, resource depletion, violent weather, growing economic and political instability, and the threat of war and conflict. It is a time that has been building for a long period. It is not merely the consequences of events today or tomorrow. It has been building over a long period of time.

Humanity has misused its natural inheritance. It has misused the world. It has overused its resources. In the name of growth and expansion, it has exceeded the limits of many of the world's resources. It has done this at the expense of your future and your future stability.

The difficulties ahead will require immense change in human perception and behavior. This will not be merely a matter of perspective or philosophy, but really of necessity. It will be a time of grave difficulty, especially for the poor peoples of the world, for they will be the first to suffer, and the cost of humanity's misuse of the world will fall most heavily upon them, even though they are the least responsible for producing these Great Waves of change.

The Great Waves of change are not merely one event, but the convergence of many powerful forces. It is their convergence that will create instability and unpredictable change and will create so many

PREPARING FOR THE GREAT WAVES OF CHANGE

interacting forces that it will be very difficult for people to understand what is happening, and why all of this is occurring.

It is dangerous because it will produce immense upheaval and instability, and people are not prepared for this. Their governments have not prepared them. Their religious institutions have not prepared them. The media has not prepared them. And so they have very little emotional or psychological preparation, and very little practical preparation for the great difficulties to come.

It is a dangerous time also because humanity is facing limitations in its resources, and there will be great competition and the risk of conflict over the remaining resources. Nations and groups will be competing with each other very intensely, and this can spill over into conflict and war on a very large scale, and in many parts of the world simultaneously.

Life teaches you the lessons of exceeding your natural inheritance. It also teaches you that nature is unmerciful to the unprepared. Your history has taught you that when you exceed the limits of what your environment can provide, it leads to great difficulty.

It is necessary for you to look into these matters with as much objectivity as possible. Your mind will want to deny it. Your mind will want to argue against it. You will want to cast blame on certain individuals or institutions. But really the Great Waves of change are the consummate result of humanity's impact upon the world and the impact of nations upon one another. Everyone is involved in this to some degree.

People believe the world will just keep providing endlessly, and when they have exhausted the world, they will go out into space to find

NAVIGATING THE DIFFICULT TIMES AHEAD

what they need. This is a reckless and irresponsible approach to life. It is heedless and it leads to disaster.

To live within the bounds and the limits of this world is the kind of stability that humanity will have to establish. And given the enormous size of the human population today, it will be an immense challenge.

It is a challenge not just to meet certain specific needs, but it is a challenge really to preserve and protect human civilization. It is really a calling to protect the human family and to establish over time a greater degree of stability and security.

The emphasis on growth and expansion must transcend to an emphasis on stability and security. This is what other nations in the universe have had to establish, for growth and expansion beyond certain limits lead to great danger and the risk of collapse.

You have lived in a time of great expansion. You embrace the idea of endless opportunities. And for a few people in the world who can enjoy it, there is great freedom here. But you are now entering a different era in a different set of circumstances.

This is not merely an intellectual understanding or an intellectual issue. You can feel it. People everywhere are feeling anxiety about the future and about the state of the world.

Here you must search your feelings. Perhaps your beliefs will be challenged by this reality. Perhaps you will deny it on this basis. But search your feelings. Perhaps you insist upon being optimistic, thinking that human technology and ingenuity will meet every challenge. But here you must search your deeper feelings.

PREPARING FOR THE GREAT WAVES OF CHANGE

Do not fool yourself. There are dangerous waters ahead, more dangerous than any situation that the human family has ever had to face before. It is not simply a few problems that must be solved. It is an overall shift in human awareness, human behavior and human relationships.

People will not make this change willingly. They will not adapt peacefully. It will be required. It will be a necessity—not forced by governments, but by life itself.

The human family will have fewer resources to use in the future—limits upon food production, very great difficulties in gaining access to fresh water and severe restraints on the energy resources. Transportation networks will break down. Whole communities will collapse because there will not be the infrastructure to support them. The cost of everything will go up. There will be financial chaos beyond what governments can remediate.

These are the Great Waves of change. If you have the courage and the objectivity, you can begin to see them coming over the horizon. When you set down your self-assuring assumptions, when you set aside blaming others, you will have the opportunity to feel and see the approach of the Great Waves of change.

It will present a specter of immense tragedy and upheaval, and you will be afraid and you will be uncertain. If you are honest with yourself, you will be afraid and uncertain—at least at the outset.

When you set aside your assumptions, and your denial, and your preferences, and all the things you insist upon and face this squarely, you will see that it is bigger than any solution you can put before it because they are the Great Waves of change, and there are many of them, and they are all converging at the same time—creating

NAVIGATING THE DIFFICULT TIMES AHEAD

confusion, crosscurrents of life, each impacting one another, creating so many different forces of change that you cannot predict them all.

If you are really honest, you will have to see that you do not have an answer, that your intellect alone cannot solve everything here. And if you are honest and allow your feelings to emerge, you will see that already you sense the approach of great change in the world.

The mind does not want to suffer, so it creates convenient solutions, convenient conclusions and simple explanations. It casts blame upon others. It asserts that technology will solve this, that humanity has always solved its problems, and it will solve all of these problems, and you try to keep the experience away from yourself.

This is understandable, but it is unfortunate because it disengages you from recognizing the great change that is even coming into your life. And it prevents you from preparing for the impacts of the Great Waves of change.

There are two realities here you must understand. The first is that the Great Waves are coming. The more you can see them and feel them and recognize them, the more you will be able to be prepared for their impacts and to place your life out of harm's way, to put yourself in the best position to navigate the difficult times ahead.

The second truth is that there is a deeper Knowledge within you, a Knowledge that God has placed within you to guide you, to protect you, and to lead you to your greater accomplishments in life. While your mind will be frightened and uncertain, Knowledge within you will emerge.

PREPARING FOR THE GREAT WAVES OF CHANGE

It is not afraid of the Great Waves of change. It is not afraid of the world. It is beyond the persuasions, the corruption and the seductions of the world. It is the only part of you that is truly reliable and wise.

It is God's great endowment. You may pray urgently to God for salvation, for guidance, for redemption, for protection, but God has already given you a perfect guiding Intelligence. Whether you are a part of a religion or not, whether you practice a faith tradition or not, wherever you live, whatever your culture, whatever your financial position, you have the great endowment living within you.

The Great Waves of change will overcome your solutions. They will be greater than your ideas. They will undermine your confidence in governments, in religion and in technology. It will be a great crisis of confidence.

But Knowledge within you is not dissuaded by any of these things. It has the Power of God behind it. It alone knows how you will be able to navigate the difficult and uncertain times ahead.

Even if the world goes mad, Knowledge within you will keep you certain and focused, alive and aware. Even if you are required to make a hundred different unanticipated decisions in the face of the Great Waves, Knowledge within you will enable you to make each one with wisdom and clarity.

You do not realize the greater power that resides beyond the reach and the realm of the intellect. You do not see its central importance to the welfare and the direction of your life, even under normal circumstances. But now you are facing unprecedented circumstances, change on a level that the human family has never had to face before.

NAVIGATING THE DIFFICULT TIMES AHEAD

You are also facing the presence of forces from the universe, physical beings who are here to take advantage of a weak and divided humanity. These small groups of forces are not military in nature. They represent resource explorers and economic collectives. They will seek to further human conflict and gain influence in the halls of power in government, religion and commerce. They represent a great danger to humanity. They will pose as enlightened beings, as benign and beneficial to the needs of humanity. But it is really a deception.

They come into the world now at a time when the Great Waves of change are building. They know that humanity is unprepared and that it will become further fractured and contentious within these Great Waves. It sees an opportunity to position itself and to build its base of power here through clever manipulation and through genetic interference with the human family.

It is a reality that is only seen by a few, and yet it poses the greatest danger to your freedom you have ever encountered. These forces are small, but persuasive. And because their intentions remain hidden, humanity is unaware of them. They are part of the Great Waves of change.

Do not think that the Great Waves of change are beneficial in themselves. Do not think that it is a necessary and valuable experience for the human family. For the Great Waves themselves are extremely dangerous, and the risk of human suffering is greater than any war that has ever been waged in this world. And the potential for the loss of human freedom and human unity is immense.

It is how you respond to the Great Waves of change that will determine whether you can gain benefit from them or not. But they within themselves are not beneficial. Do not fool yourself in thinking that they are, or you will make a grave and critical error.

PREPARING FOR THE GREAT WAVES OF CHANGE

Humanity will need great assistance. Alone you will not have all the answers. And where will you find the strength to change your views and your behavior and to prepare your lives for the impacts of the Great Waves of change? Where will you find the strength to overcome your resident tendencies and the potential for conflict between people, groups and nations? Where will you find the courage to face the Great Waves of change? And what can reveal to you the real impacts and the real nature of these Great Waves of change, including the Intervention that is occurring in the world by forces from the universe?

Only God can prepare you for these, the greatest events in the history of mankind. And God has sent a New Revelation into the world for this purpose, for the protection and advancement of humanity. It is built upon the reality of all God's Revelations.

This is difficult to understand because generations of people have changed God's Revelations to accommodate commercial, political and social realities, so it is very difficult to recognize that this New Revelation is actually connected to all the previous Revelations because the previous Revelations are so greatly misunderstood and have been compromised to a great degree by so many different forces.

Only God knows how to prepare you for what is to come. Only God knows how to prepare you for the realities of life in the universe, which you will now have to face because of the Intervention that is occurring in this world. Only God knows what humanity will have to do and to change and to establish to acquire real stability and security in a changing world.

You yourself cannot figure all this out. No one can. There will have to be a thousand solutions brought to bear. But even here, where will people find the strength, the commitment, the wisdom and

NAVIGATING THE DIFFICULT TIMES AHEAD

the courage to change what must be changed, to establish a new unity within the human family, to assure the distribution of needed resources and to avoid the great temptation of conflict and war?

Only God knows. But God has placed Knowledge within each person. So there is a possibility that Knowledge may be stimulated sufficiently that people will act in their own best interests.

You have created a set of problems over time through humanity's abuse and misuse of the world that now are bringing with them great and grave consequences. The impact on your climate and environment is so significant that it could reduce significantly the world's production of food. And the loss and the limitations of your energy resources could change and impact the lives of everyone in the world, especially in your rich nations. Do not think that your wealth will protect you now, for much of your wealth will evaporate and will be lost.

Your mind will contend against this, of course, but in your heart you know that great change is coming. This is speaking to the truth at a deeper level, beyond the level of human ideas, preferences and beliefs.

Do not think you can run away from the Great Waves of change, that you can move out into the country or find some safe place where the impacts will not affect you. This is really a grave mistake because there is nowhere where you will be free from these impacts, and the further you are away from the centers of human wealth and distribution, the more imperiled your life will be. This is the danger of isolation.

There will be many false leaders that will arise in these difficult times. Bad advice will be plentiful. There will be many movements to condemn other peoples. There will be claims of salvation that will be false. Many selfish and deranged individuals will seek to rise to power under these turbulent circumstances. It is a time of great risk for the human family.

God knows that humanity cannot prepare itself and has not prepared itself adequately for such great and tumultuous events. That is why a New Message has been sent into the world—to teach The Way of Knowledge, to stimulate Knowledge within people, to provide clarity about what is coming over the horizon and to prepare humanity for its future and its destiny within a Greater Community of intelligent life in the universe.

It is a Message unlike any that has ever been sent to the human family because humanity is facing such unprecedented events. While human conflict and deprivation and environmental collapse have been a part of humanity's history, you have never had to face a set of circumstances such as you are facing now. This will not only affect certain groups in certain places, but will impact the whole world— the entire economy of the world, the political infrastructure of the world, relations between nations in the world. All will be dramatically impacted by this.

People are shortsighted of course. They attribute change to immediate events or to theories. They think it is just an economic cycle, or it is just a few problems that have to be solved. They think that life will go on as they have known it.

Because of this, they do not see, and they do not know what is coming over the horizon. And because they do not see and do not know, they are not prepared. They are not prepared as individuals. They are not prepared as groups and as nations and so forth. Still clamoring for ever-greater wealth, for ever more resources, for ever

NAVIGATING THE DIFFICULT TIMES AHEAD

more power, they are hastening and furthering the onset of the Great Waves of change.

It is as if humanity were going to be bombarded by a thousand things at once. People are aware of a few of these things, but cannot see the whole picture. That is why the Creator of all life has sent a New Message into the world to prepare humanity for the Great Waves of change and to prepare humanity for its encounter with the Greater Community of life in the universe. These are the two greatest thresholds the human family has ever had to face, greater than any problem that you can identify in the world today.

People think small. God thinks big. God's Message is really big. It is very inclusive. It covers many things. It sets out a preparation that people everywhere can use to begin to prepare their lives and themselves for living in a world of turbulence and uncertainty.

This, of course, exposes the whole range of human assumptions that insists upon all of the forces that are leading humanity astray, and all of the assumptions that human will and human diplomacy are sufficient to deal with whatever problems may arise in the face of these Great Waves of change.

But your deeper feelings tell you something different. If you can feel them, if you are aware of them, they will tell you something different, for they are responding to the world. And the world today is giving you signs of what is to come. And Knowledge within you is giving you signs of what is to come.

Knowledge within you is speaking to you every day, but you cannot hear it because you are up at the surface of your mind, in your worldly mind—a mind that has been conditioned by your culture and your family, your beliefs, your fears, and your desires.

PREPARING FOR THE GREAT WAVES OF CHANGE

But deeper down within you, there is a greater Mind, and this deeper Mind is not afraid. It is not condemning. It does not generate hate and blame. It does not vacillate. It is not unstable. This is the great compass that God has given to you. And it has been given to you especially because you have come into the world at a time when you would experience the Great Waves of change.

The requirements upon you are greater than your ancestors. The preparation is different. What Knowledge has been equipped to provide is greater and different and unique to the world that you are facing today and will face in the days and years to come.

You are sent here for a purpose. You are sent here to face the Great Waves of change and to contribute something special, unique and beneficial to the human family. You have been sent to build a new future, a better future, a future that will be unlike the past. And you have not come alone, for there are others who are destined to help you to discover this greater purpose and to fulfill it in life.

Perhaps you have not yet met these people, or perhaps you have met a few of them already if you are very fortunate. But they represent a greater life and a greater intention that Knowledge within you represents.

Knowledge will stimulate human innovation, human invention, human collaboration, and human unity. It will inspire new technology. It will inspire greater creativity. It will inspire greater cooperation. All the things that human leadership and citizens everywhere will have to establish and adapt to that are truly beneficial will be inspired by Knowledge. This is why what is invisible and mysterious within you has the greatest power to effect change in human awareness and behavior.

NAVIGATING THE DIFFICULT TIMES AHEAD

Do not think that simply building new technology alone [will work] because you may not have the energy to build it. You may not have the economic stability to build it. You may not have the resources to build it. People take all these things for granted. They think there is an endless supply. But they do not realize that in the face of the Great Waves of change all of these things will be challenged.

You may at times feel helpless and hopeless. But you have Knowledge within you, and it is not helpless and hopeless. Whatever your individual strengths and weaknesses may be, no matter what your particular circumstances offer you at this moment, you have this power to guide you. That is why taking the Steps to Knowledge is so very important, particularly at this time.

Your faith here cannot be blind. Do not think that God will just intervene and fix everything for humanity, for this does not happen in life. Do not think you will be saved in the last hour by some miraculous intervention. Do not think that races from the universe will come and rescue humanity, without taking from you what they want.

This is a problem for humanity. It is a problem the human family must solve. You cannot be helpless and hopeless or have a welfare mentality regarding this. Take stock of your strengths. Take stock of your weaknesses.

God's New Message has provided a set of guidelines to help you prepare for the difficult times ahead. These are general and are meant to give you a start. But there is no set of instructions that would be able to speak to everyone's circumstances and needs.

Your first responsibility is to face the Great Waves of change. If you cannot do this or will not do this, then you cannot prepare. And

PREPARING FOR THE GREAT WAVES OF CHANGE

when the Great Waves strike, you will be overcome and overtaken without the necessary resources to deal with this.

This is a time now to stop blaming others, to end your ceaseless debates, to set aside your grievances and your hostilities, and to face the fact that you are going to have to live life much more simply in the future. You will be living in a world of great resource restraints. You will not be able to get around very easily. Everything will be very expensive, and there will have to be a great emphasis on producing food and other necessary things on a much more local scale.

There will have to be tremendous restraint of anger, blame and hostility if people are to make this necessary transition. So many assumptions and beliefs will have to be set aside to face the reality of your situation.

The guidelines that the New Message has provided will give you a beginning start to position your life more securely and more safely, and to provide you the emphasis that you will need to reconsider your interests, your actions, your commitments, your involvements, how you spend your wealth, how you are going to remain employed and how you are going to be able to be stable while people around you become unstable. The guidelines are here to help you maintain clarity, stability and self-determination in the face of ever-growing chaotic events around you.

The world will seem like it has gone mad, but you cannot become mad. Within you is a greater compass and a greater strength. You have always needed this, and you have suffered immensely because you have not experienced it sufficiently in the past. Now you will need it to survive, and to take care of your families and your loved ones, and ultimately to provide a greater service and stability to your communities, and your nation, and the world as a whole.

NAVIGATING THE DIFFICULT TIMES AHEAD

Here humanity cannot use weapons and force to achieve its goals. It must use wisdom, clarity and cooperation. This will be driven by necessity, not by ideology, not by good ideas or a sense of philanthropy. It will be guided and determined by necessity.

Unite, and you will survive and can advance. Yet if you fight with one another, you will decline and forces from the universe will gain greater strength here—a strength that will eventually overtake human sovereignty in this world.

The challenge before you, then, is immense and unavoidable. It is what you can see now and do now that will make all the difference for you. You are not in a position to save humanity, but you are in a position to secure your position and your life, to stabilize your life, to prepare your life, and to assist others in preparing their lives.

For it is your communities that must become strong. It is your communities that must adapt to changing circumstances. You can work effectively on a local scale, and you will have sufficient challenge in all directions to keep you engaged in a productive way.

The stronger you are with Knowledge, the more you will be a force of inspiration for others, and the more creative your mind will be in seeing ways in which you can stabilize your life and be of service to others. For it will not work if you secure your life alone, only to have your community fail. Here your preparation cannot merely be a selfish pursuit. It will have to support the well-being of people around you.

There will be whole segments of your population that will be greatly at risk—the elderly, single parents with children, people who are disabled. These especially will be at risk, and you must have the strength to care for individuals here as well as your own family.

PREPARING FOR THE GREAT WAVES OF CHANGE

There will have to be incredible human contribution and service if humanity is to endure and survive the Great Waves of change.

While the Great Waves of change represent an immense danger and tragedy for humanity, there is also a redemptive power to them—if you can respond to them appropriately. They can bring clarity and purpose to your life. They will require selfless actions and a deeper commitment to your well-being and the well-being of others.

The Great Waves of change will draw you away from personal addictions and self-obsession, requiring of you greater service to the people around you. They will break your attachment to wealth and pleasures that are unhealthy for you. They will give you purpose, meaning and direction. They will organize your mind, your actions, your thinking, your behavior, and the beneficial use of whatever resources you have at hand.

The Great Waves of change are not beneficial within themselves, but they can become beneficial, depending on how you respond to them. After you get over the initial shock and dismay at recognizing what is really coming over the world, then you have to get to work in preparing your life and helping others to become aware and to prepare their lives. This gives you greater purpose, meaning and direction.

Whatever is holding you back in your life, whatever form of self-obsession or self-conflict is holding you back will have to be overcome out of necessity. You will have to either sink or swim. And you must swim. In this, there is a power of redemption.

Here Knowledge within people becomes stimulated and strengthened. A deeper conscience within people becomes activated. Here people collaborate, bringing their unique strengths and talents

NAVIGATING THE DIFFICULT TIMES AHEAD

to bear to solve the innumerable problems that the Great Waves will generate.

And you have a New Message from God in the world to prepare you, to alert you, to empower you, and to empower all the world's religions to be of the greatest service and benefit to humanity at this great turning point in your evolution.

Therefore, do not shrink from the Great Waves of change. Do not deny what is coming over the horizon. Do not fool yourself into thinking that it is a problem that can be solved in any way that you can understand presently. Do not condemn and blame others, for everyone has created the Great Waves of change.

Instead, recognize them, face them, pass through the initial phase of shock and dismay, and then get serious about your life. Take the Steps to Knowledge. Become stronger with Knowledge. Feel it moving within yourself every day. It will give you clarity and sobriety. It will take you out of foolish behavior and foolish thinking. It will set you on a real path of preparation and will give you greater strength and certainty than anything else in the world can provide.

While institutions falter, while governments seem incapable of responding, while people around you go into denial and despair, you must have this greater strength and greater objectivity within yourself. Take stock of your feelings. Look out over the horizon of the world and ask yourself: "What is really coming? What must I prepare myself for? Who do I know can help me prepare? Who in my life is strong, and who in my life is weak? What must I build now to prepare myself for the future? And what must I let go of or avoid to be strong enough to face the future?"

PREPARING FOR THE GREAT WAVES OF CHANGE

These questions are part of a deeper evaluation that the New Message from God is recommending. It is providing the context and the questions that must be asked at this time while you have the freedom to ask them—before the Great Waves are crashing upon the shore, before nations and groups are overtaken. It is in this time of quietude and relative peace that you have the greatest opportunity to become aware and to begin your preparation and the deeper evaluation of your life that will be required.

God has given you the power and the presence of Knowledge. It is beyond the range and the realm of your intellect, but it will employ your intellect and all of your talents and skills for a greater service and a greater good in the world. This has always been your promise, but now you are facing the Great Waves of change, and they will require these things of you. They will require a greater commitment, a greater wisdom and a greater courage.

Do not look to leaders of institutions or government to do these things for you, for this is a personal calling for each person. Do not simply rely upon others to take care of these problems for you, for if you do, your life will be unprepared, and you will find to your great dismay that those leaders who you relied upon will fail your expectations.

The Great Waves will either seem a crushing defeat or a real call to human power and strength, depending upon how you respond and what motivates you. Are you motivated by fear and personal preference or by the power and the strength of Knowledge that God has put within you to prepare you and to guide you for these very events that you are destined to face?

The choice is yours, and it is a choice you will have to make continuously. But you have the power to make this decision. And

NAVIGATING THE DIFFICULT TIMES AHEAD

while the world is still calm, before the Great Waves really begin to strike, now is the time to look, to see, to know, and to begin your preparation and your deep evaluation of your life.

For the truth is you might have to be in a very different set of circumstances, even engaged with different people, to really be in the position you must be in to navigate the difficult times ahead. Do not think that you will have to stay necessarily where you are, doing what you are doing now. For there may have to be great change in your circumstances. Have the courage to face this and to consider this. Do not hold on to what little you have, for everything must be re-evaluated now.

You must build your shelter before the rains come. You must have adequate resources before the winter arrives. Your life must be in the position that Knowledge will guide you to establish before the Great Waves of change really begin to impact humanity. For when that happens, your choices will be limited, and it will be far more difficult and in some cases impossible to reposition your life.

People who have vision can see what is coming, where others do not see. They can listen to the power of Knowledge within themselves, where others do not hear. They will take actions to prepare for coming events, where others will not see the need for doing this.

That is the power and the presence of Knowledge within you. This is your greatest strength and your greatest gift to others. For this is why you have come into the world at this great turning point.

As revealed to Marshall Vian Summers, September 29, 2008

CHAPTER 2

Facing a World in Decline

You are facing a world in decline, a world whose resources have been overused and misused by the human family, out of greed, corruption, ignorance and presumption. It is a world that has been stressed, a world whose supportive capacity for the human family has been overtaxed and overused.

There are growing numbers of people who are aware that there is a serious problem here, but governments are not responding, and the wealthier people of the world do not want to question the source of their wealth. They do not want to pay the expenses that will be required to mitigate the damage that has been placed upon this world.

You are facing a world in decline, a world of declining resources, a world whose climate is being affected now and is being changed, leading to violent and unpredictable weather and the loss of water resources, critical resources that are needed for large populations of people. You are facing a world where a growing humanity will have to drink from a slowly shrinking well.

This is the world that you have come to serve. Do not deny it. Do not avoid it. Do not be dishonest regarding it. For it is the world you have come to serve.

PREPARING FOR THE GREAT WAVES OF CHANGE

The circumstances that are arising now—the Great Waves of change coming to the world—have been foreseen. God knows they are coming. For those who have great vision, they can see the Great Waves out on the horizon. Perhaps they cannot interpret them precisely, but they can see them out on the horizon.

The human family will have to face now a set of circumstances that as a whole it has never had to face before. And yet so few people can see this. It is one of humanity's weaknesses that it does not look ahead.

Believing that the world is an endless cornucopia of resources, humanity is forging ahead with its endless emphasis on growth and expansion. It is like an addict that cannot stop himself from deepening his addiction, the human family has not gained yet the cohesion and the cooperation necessary to adapt to these changing circumstances and to the great trials that await it.

There are several things here you must face. You must face that you are living now in a world in decline. You must go through the shock and the awe of recognizing this and attain a more powerful state of clarity, courage and determination. You must face this now and look at the evidence that is mounting.

This is necessary to give you encouragement and the sense of responsibility to prepare—to prepare for a world in decline, to reassess your position and your circumstances, to see where you are strong and where you are vulnerable.

This deep evaluation is necessary now, for the world is changing. Your environment is changing—your physical environment, the social environment, your economic environment. It is all changing now.

FACING A WORLD IN DECLINE

Those who will be able to navigate the difficult times ahead will be those who have looked and seen and have prepared themselves. Do not wait for a consensus with others here, or it will be too late. Then you will have very few options, and the choices will be immensely difficult.

Next, you must recognize that you have come into the world to live in these times, to face the Great Waves of change, to adapt to a world in decline, and within the many needs of humanity, to see where your contribution is most essential. This indeed is your time. These are the circumstances that will call out of you your greater gifts, so do not reject them. Do not deny them. Do not run away from them, for they hold the power to call out of you your gifts and to initiate your redemption here.

A third thing is you must reconsider your demands and expectations upon the world. Beyond meeting fundamental needs here, what do you expect or desire from the world? What do you want the world to provide for you beyond your basic needs? Endless possessions? Endless opportunities? The ability to support you in whatever you want to do and having whatever you want to have? What are your expectations here? Can these expectations be met? And are they ethical?

How you behave, what you consume—your standards here are very important because what will tip the scale in humanity's favor are clarity, courage and determination. You must have clarity about what is coming and the great challenges now facing humanity. You must have courage to face this great threshold and to face and to accept what it will call forth from you as a gift and a contribution to humanity.

PREPARING FOR THE GREAT WAVES OF CHANGE

And you must have determination because part of you will not want to face this. The weaker part of you will not want to face this or accept it. It will want to go into denial so that you can preserve your personal goals and fantasies. You will not want to face its challenge, its requirements and its opportunities.

People around you will remain blind and ignorant, their backs to reality. And when the Great Waves come, they will be caught off guard. They will be enraged. And they will be terrified.

You must have determination while others are remaining ignorant. You must have determination while others are in a state of denial. You must have determination when others give up, feel hopeless and helpless. You cannot succumb to this.

This requires great sobriety and a great love for the world and the willingness to look and see what is coming over the horizon. If you can see, it will be tremendous, and you will feel overwhelmed. Perhaps you may go through a period where you feel hopeless, unable to see a remedy. You will look at the condition of humanity's leaders and the general public and you will wonder how anything can ever be accomplished, and how any preparation can be undertaken on a large scale. It can appear hopeless, indeed, and the chances can seem to be very slim.

But what is important here is that you follow Knowledge within yourself, the deeper Intelligence that the Creator of all life has given you. It is not afraid. It is not corrupted. It is not compromised. As a result, it remains as a source of your strength, your integrity and your contribution to a world in need.

Knowledge functions beyond the realm of the intellect. You cannot figure it out, but you can experience it. And you can receive its

FACING A WORLD IN DECLINE

guidance, its restraint and its motivation. With Knowledge, you can navigate what seems to be unpredictable and hopeless to achieve a greater result.

Here you must participate even if you have no idea how it is going to work. And you must function against the odds, even against all hope. Knowledge is not guided by fear and hope, ambition or desperation. This power is so formidable in the face of changing circumstances that you really cannot find anything else in life that will give you this strength, this certainty and this stability.

But you cannot be passive and rely upon Knowledge, for it will require that you be observant, powerful and responsive. You must realize the great danger, but Knowledge will give you the power and the encouragement because it is a gift from God.

You must not fall prey to the persuasions of hopelessness and self-doubt. These will be real challenges for you if you are to face the Great Waves of change—environmental destruction, changing climate and violent weather, growing political and economic instability, diminishing resources, and the growing risk of competition and conflict between nations and groups over who will have access to the remaining resources.

These are the Great Waves of change. And because they are all converging at the same time, they create a powerful set of forces, which influence one another in ways that are almost impossible to predict.

Never think that if the world is depleted, you can go into space to find what you have lost here on Earth. For beyond this solar system, resources are owned by others, by other nations, and they are far more powerful than you are. Never think that the universe is a big, empty place awaiting human exploration and exploitation. For there are societies in your vicinity of space that are far older than any civilization here on Earth, and they will not countenance intervention or invasion from any human group.

What this means is that you have to make it work here on Earth. If you cannot do this, you will fall under the persuasion of foreign powers in the universe. Some of these powers are even here in the world today, seeking to influence humanity for [their] own purposes. Your declining prospects here give [them] a great advantage of persuasion, for these forces are not military in nature. They are here to take advantage of the world's resources—its natural resources and its strategic position in this part of space.

God knows that you cannot see all of this on your own, and that is why a New Message has been sent into the world for the protection and the advancement of humanity. There is a New Message from God in the world. It is here to warn, to strengthen and to prepare humanity for the Great Waves of change and for the immense challenges of encountering a Greater Community of intelligent life in the universe.

These two great challenges, more than anything else, will shape the future and the destiny of humanity. Each has the power to undermine and destroy human civilization. And each can be countered by the power and the presence of Knowledge within people.

It is a great challenge indeed, greater than anything the human family as a whole has ever had to face—greater than any world war, greater than any famine or pestilence, greater than any economic tribulation.

The Great Waves of change and humanity's encounter with competition from beyond the world will, more than anything else, give humanity its greatest opportunity for human unity and cooperation, for the protection of humanity and for the welfare and sustainability of life here in this world.

Do not think that technology alone will solve these great problems. For without human unity, courage, clarity and determination, technology alone will not be enough. And you will be seduced by races who are more technologically advanced than you.

You must look ahead. You must see the signs of the Great Waves of change. And you must see the immense evidence of foreign presences here in the world. You must do this without assumptions, without preferences, without trying to use all this to fortify your preconceived ideas and your ideology. You must look to see and to learn. It sounds simple, of course, but it is difficult for people to do this.

Perhaps in recognizing these two great realities, your ideas will be challenged; your assumptions will prove to be wrong. Have the courage and the humility to face this. Have the honesty to recognize that your education must begin anew.

The New Message from God will provide the missing pieces that humanity could not provide for itself regarding what is coming over the horizon and the realities of life in the universe. There are limits to what humanity can see on its own, and that is why this great gift from the Creator of all life is so very important.

Here you must not be guided by hope or by fear, but instead by the power and the presence of Knowledge. This will enable you to be clear and objective and able to discern the circumstances as they emerge.

PREPARING FOR THE GREAT WAVES OF CHANGE

Complications will arise. Difficulties will arise. Unexpected and unanticipated events will occur. If your mind is still and you are listening deeply within yourself, you will know what to do. You have a greater set of strengths that are not cultivated and have not been called upon sufficiently for you to have this clarity of mind. But that is all changing now.

Humanity is now facing a more difficult world, a world that will require greater cooperation and greater selflessness and greater contribution from ever-growing numbers of people. A divided, contentious humanity will not succeed in the face of the Great Waves of change, and can be easily manipulated by clever foreign powers who are already in the world seeking to take advantage of a weak and divided humanity.

Look into these two great realities, and let them inform you what you must do. Learn to take the Steps to Knowledge to still your mind so that you can listen to the deeper Mind within you, the Mind of Knowledge.

Here the inner preparation will be greater than the outer preparation, for if you cannot find this position of clarity, courage and determination, then nothing you do on the outside will be successful. You can make critical mistakes in judgment, and you can follow people who will lead you into greater danger and vulnerability.

Facing a world in decline, your mind will be afraid. It will want to go into denial. It will complain. It will project blame. It will seek for answers and solutions, not wanting to have to face the uncertainty of living with a greater problem.

Here the degree of your education does not really matter. It is the quality of your education that matters. Even your most highly educated scientists and leaders are in complete denial and ignorance of the Great Waves of change. They blithely think that whatever it is, humanity can deal with it. "No problem. We'll create the technology. We'll take care of that problem when it emerges," they think.

These are the assumptions of weak and foolish minds. They think that humanity cannot fail, and so they think the problem is not that significant. But the problems are that significant, and humanity can fail.

Civilizations have failed before, leaving only a few traces. Do not think that human civilization currently cannot fail, cannot be undermined by the very forces—the forces of resource depletion, the forces of the loss of food production, forces of political instability and human conflict, and forces of Intervention from races from beyond. All of these have led to the collapse of previous civilizations. Do not think that because you live in a somewhat modern life that the rules of nature do not apply to you.

The Great Waves of change have the power to undermine human civilization. And those forces who are intervening in the world today, who are deceptive, will seek to undermine human civilization for their own benefit. They know that humanity is moving into a position of immense difficulty and vulnerability in facing the Great Waves of change. And they will project themselves, these intervening forces, as the saviors of humanity. It is a perfect environment for intervention. Because there are other races who covet this world, these attempts will be made repeatedly.

This is a giant wake-up call for the human family, of course. It is redemptive in that it brings you back to yourself, to your senses. And it calls upon the deeper Knowledge that God has placed within each person to guide them, to protect them and to lead them to a greater life of contribution—a life grounded in reality, a life that is inspired by the Power and the Presence of God.

Whether you are religious or not, this is your great opportunity. No matter what nation or faith tradition you have come from, this is your great opportunity. It is driven by necessity now. It is not a casual choice. It is not only for the elite. It is for everybody to respond.

Receive, then, the New Message from God, which alone has the wisdom and the power to educate humanity about the Great Waves of change and the reality and the spirituality of a Greater Community of intelligent life in the universe. Take the Steps to Knowledge to learn of the deeper Intelligence within you and to receive its guidance, its restraint and its inspiration. Look out on the horizon of the world and see what is coming. And plan your life around what you see, not around what you want or what you prefer.

This is calling upon a greater strength within you, a strength that you have rarely known, a strength that exists beyond the realm of your intellect, your beliefs and your ideology. You cannot be fooling around in the face of the Great Waves of change. You cannot be self-obsessed and blind and unresponsive to the environment and be in a position to deal with the power and the powerful influence of those races who are intervening in the world today.

Humanity can largely mitigate and adapt to the Great Waves of change. And humanity can throw off Intervention and establish its own rules of engagement regarding all who might come to the world, both now and in the future. You have this power and this responsibility. Fail to claim it, and nothing will save the human family.

FACING A WORLD IN DECLINE

The choices are so fundamental and so essential here that they cannot be avoided. An endless debate about the seriousness of the great environmental changes to come is such a massive waste of human energy and opportunity. These debates are carried on without Knowledge, for the mind endlessly speculates and tends to perceive reality according to its preferences and its prejudices. That is why you must follow what you most deeply know and not simply call upon experts to guide you and to educate you, for they may not know.

Humanity can survive in a declining world, but it will be a very great challenge. You will have to control both population and consumption. This seems to be unacceptable to freedom-loving peoples, but freedom requires responsibility. Freedom is a privilege and not a right. It must be earned. It must be protected. It must be guarded. And it must have an environment that can sustain it.

Humanity must outgrow its adolescent phase of growth and expansion, for you are running out of room. You are running out of resources. You are reaching the limits, and in some cases have surpassed the limits, of what this world can sustain.

That is why this calls for a deep evaluation, a reconsideration of your life and priorities and a reassessment of humanity's future and the forces that will impact it. Knowledge will guide you to make the real contribution you are here to make. It will bring into your life the essential people that will make the discovery and the expression of this contribution possible and meaningful.

Therefore, your challenge is both on the outside and on the inside. Your first responsibility here is to build a connection to Knowledge and, while you are doing this, to undergo the deep evaluation regarding your life and circumstances.

PREPARING FOR THE GREAT WAVES OF CHANGE

Here everything must be put up for reconsideration—where you live, how you live, how you travel, the people you are with, the quality and strength of your relationships, an assessment of your skills and weaknesses. And this must lead to a plan. Perhaps this plan will not be complete at first, but it must be a starting point. For you must move off of the beach when the Great Waves are coming.

It will take time to change your life and circumstances. You need this time. Do not delay. The more you delay, the fewer are your options. You must be prepared to act decisively and courageously while others are seeming to do nothing. You must follow Knowledge, not consensus. This will prove to be pivotal in your ability to prepare and to navigate the uncertain times ahead.

It is you who must not become a victim of life's changing circumstances, but its beneficiary. It is you who must be in a position to assist others in preparing for the great change that is to come and for the challenges of emerging into a Greater Community of intelligent life in the universe.

Time is of the essence, and preparation is the key. God is providing now the vision of the future that you must have. And God is providing the preparation in Steps to Knowledge, and in the prophetic teaching about the Great Waves of change and the challenge of emerging into a Greater Community of intelligent life in the universe.

This New Message presents what is essential for you to know and for humanity to know—to understand what is coming, to see it without deception and distortion and to begin the preparation that must happen on both an individual and collective level.

FACING A WORLD IN DECLINE

In facing the Great Waves of change, people will have to leave desert regions. They will have to leave areas that are prone to flooding. A great emphasis will have to be placed upon agriculture and water resources and fuel resources. There will be immense human migrations, refugees forced to leave their homelands because the deserts will be growing and the coastlines will be endangered by violent weather and rising waters.

You can see this even now—if you become objective, if you study, if you become educated and if you listen to Knowledge within yourself.

Entire nations will collapse in the face of the Great Waves of change, creating immense instability in their regions. The human need will be unprecedented and immense. The need for human contribution will be tremendous, and the wealthy must dedicate their resources to meeting these needs.

There will have to be a cessation of war and conflict, particularly between large and more developed nations. Energy resources will have to be managed wisely and will have to be conserved, even to extreme extents.

Do not wait and think that humanity can deal with these problems in its own time. For time is of the essence. And people tend to greatly underestimate the great changes that are coming to the world. It is this inability to look forward and to prepare and to moderate one's life that represents humanity's vulnerability.

And it is the assumption that you are alone in the universe, or that no one can reach your shores, or that whoever would come here would come here to benefit humanity—these represent your weaknesses in considering the reality and the consequences of contact with intelligent life in the universe.

PREPARING FOR THE GREAT WAVES OF CHANGE

Part of your deep evaluation is to determine your strengths and weaknesses with as much objectivity as you can and to determine humanity's strengths and weaknesses in the same regard.

You can do this because the power and presence of Knowledge is with you, because God has equipped you to live in a declining world. Knowledge within you is not afraid. It has immense confidence. You too must have this confidence if you are to fulfill your destiny in this world and if you are to build a new foundation for humanity, to build a new future—a future that will be unlike the past in a world that will be unlike the past.

This is your destiny and this is your calling. It will give you the greatest satisfaction and fulfillment to meet this great challenge and to realize your greater strength and your unique gifts that are meant to be given at this time.

As revealed to Marshall Vian Summers, October 30, 2008

CHAPTER 3

Living in a Time of Uncertainty and Instability

It is evident that you are entering into a time of great instability and uncertainty in the world, a time where the Great Waves of change are beginning to strike the world: resource depletion, environmental degradation, violent weather, growing economic and political instability and the risk of war and conflict over the world's remaining resources.

It is a time that will represent the results of humanity's misuse and overuse of the world. It is a time where there will be so many currents and crosscurrents that it will be impossible to predict exactly how things will turn out.

Part of this uncertainty is how people will respond to the Great Waves of change and to their own trials they will be facing within their circumstances. This will affect everyone in the world. It will be most difficult at the outset for the poorer peoples of the world, who will struggle to survive, who will face increasing pressure—economic pressure, social pressure—as the cost of everything goes up and as opportunities diminish.

But the effect of this will reach far into the wealthy nations, where people everywhere will be losing so much of their wealth as the result of mismanagement and incompetence in managing the financial world. But more significantly, it will be the Great Waves of change that will alter the human landscape.

Do not underestimate the power of the Great Waves. They are far more powerful than economic or political policy or the beliefs and assumptions of the masses of people.

It is a time of great reckoning now, a time of decision, a time to face great uncertainty and upheaval in the world. You will have to face this, and perhaps you are beginning to face it even now.

There is no escaping from this, no escaping into fantasy or self-obsession, no escaping into all the myriad routes of escape that human cultures have created. There is no escaping even logistically. You cannot run away to the country and be safe, for in most cases being isolated there will not be safe at all.

You will see much anger and denial, much blame and condemnation arising as people's circumstances are altered beyond their control and as their wealth and opportunities diminish in so many cases. You will see the results of people's lack of vision and preparation regarding the Great Waves of change.

They will be caught off guard. They will be unprepared. And the shock will be very difficult and, in many cases, terrible and extreme. All of a sudden they find themselves now in a changing landscape. Doors are closing to them, and they did not see it coming.

All great change in the world brings with it signs that precede it, signs and indicators. And within you, at a deeper level within you, there is the power and presence of Knowledge, a deeper Intelligence that God has put within you to guide you, to protect you and to lead you to your greater accomplishments in life.

LIVING IN A TIME OF UNCERTAINTY AND INSTABILITY

It too has been giving you signs and indicators, but most people do not respond, being preoccupied with their interests, their desires, their difficulties or their conflicts. They are too distracted to recognize either the signs of the world or the signs emanating from Knowledge within themselves.

As a result, people are caught off guard by changing circumstances in the world. They do not respond until the last minute and then it is really too late to do very much at all.

Many people will panic. Many people will turn to very dangerous forms of behavior, of action, while others will simply lose all heart and feel victimized by the very circumstances that they could have foreseen. Many other people are circumstantially so limited or handicapped that they really are not in a position to do much of anything.

The world is changing. The climate is changing. Your circumstances are undergoing a kind of shift, but without a certain outcome. So many things will come into play now that it will be very difficult to foresee the outcome. And how humanity will respond, whether it will unite to face these difficulties or whether it will fight and compete within itself, remains to be seen.

For you, however, it is time to face the reality of the Great Waves of change—to begin to evaluate your position, your circumstances, your strengths, your weaknesses, your assets and your liabilities. You cannot help others or be of much good to the world if you yourself are overwhelmed or overtaken by the great changes that are coming.

There are some things you must do right now. You must learn about the Great Waves of change. You must have the courage and the objectivity to face this, not simply to use it to fortify your existing prejudices, beliefs or attitudes, but to face it as a student would face a new education.

Next, you need to consider where you live, how you live and how you travel about to see if these things are going to be sustainable within the changing circumstances of your life—the kind of work you do, where you live, your transportation, the strength or weakness of your relationships, the strength and weakness of your health—all these things have to be re-evaluated in light of the Great Waves of change.

Yet to find the strength to do this and to find the objectivity and the clarity you will need amidst your own reactions, your own concern, your own frustration, this must come from a greater power within you, from the power of Knowledge within you. Your intellect will be overwhelmed, your emotions will be excited, your frustration will grow and increase as will your anxiety about the future.

You must awaken from your dream of self-fulfillment. You must be willing to reconsider all of your goals and objectives and the value of your relationships in light of the Great Waves of change. You cannot rely upon your own intellect, your own ideas to give you the strength and the objectivity to carry out this evaluation. The power for this must come from a greater power within you, the power of Knowledge.

Knowledge is not afraid of the world. Knowledge is not afraid of change and the possibility of loss and deprivation. Knowledge is not persuaded or manipulated by the values, the beliefs or the assumptions of others.

For Knowledge remains pure within you. It cannot be corrupted. It cannot be persuaded or manipulated by any force in the world or the universe. It represents the seat of your wisdom and the source of your

LIVING IN A TIME OF UNCERTAINTY AND INSTABILITY

integrity. Only it will have power to enable you to re-evaluate your life wisely, to see what must be altered or changed, to see what must be strengthened and enhanced.

Your mind will be in a state of confusion if you honestly face the Great Waves of change. You will have to admit that you really do not know what to do. You might think you have a plan for immediate circumstances, but things could change a hundred times in ways you cannot anticipate. So you cannot plan very far ahead, and you must question your own assumptions about what will save humanity or rescue humanity.

You are facing a world where a growing human population will have to come to a well of gradually shrinking resources. How will people deal with this? Do not think that technology alone will solve this dilemma, for technology requires the very resources that you will have an ever-greater difficulty in securing. Given the condition of people in various nations and the condition of life in your nation, how will people respond to this kind of fundamental challenge?

Do not think you have an answer, and do not put much faith in your assumptions. You have to be alert now and responsive. You have to be prepared to make important decisions and to alter the course of your life, perhaps several times, as events around you change. You have to be prepared to alter your thinking and to reconsider what you are doing.

Yet where will you find the strength and the objectivity to act wisely under such changing circumstances? And how will you be able to maintain your equanimity and clarity of mind while others around you are sinking into despair and confusion? These are important questions. Do not dismiss them. It is all part of facing a world of growing uncertainty and instability.

PREPARING FOR THE GREAT WAVES OF CHANGE

Within you there is strength and there is weakness; there is wisdom and there is folly. You must find your strength and your wisdom, and that is at the level of Knowledge—deep beneath the surface of your mind; beyond the reach of your social conditioning, preferences, beliefs and attitudes. This deeper Intelligence, which so few people in the world have ever really discovered significantly, must be your compass and your guide now.

That is why it is necessary to take the Steps to Knowledge, to connect your worldly mind with your deeper Intelligence. They must be connected because your worldly mind will need the guidance, the wisdom and the counsel of Knowledge within yourself. This is how God will speak to you.

You will find yourself being restrained and held back in many situations and motivated in others. For to Knowledge, everything is a yes or a no. You move here; you do not move here. You make this decision; you do not make this decision. You choose this over that.

To experience the power of this, put all of your questions into a form where they can easily be answered as yes or no. But you must be very honest here. Some questions are dishonest. They are choosing between two things, neither of which is appropriate. You must take things down to their essential question, their starting question. This is how you prepare your mind to engage with Knowledge.

Knowledge is not another intellect or personality within you. It is very fundamental. It is very wise. It is very powerful. You cannot pose to it trick questions or manipulated questions or ignorant questions and gain a response. It will not respond to this.

Therefore, you must ask a very basic question, the beginning question. You may ask, for example: "Should I go here or should I go

LIVING IN A TIME OF UNCERTAINTY AND INSTABILITY

there?" but that is not the first question. The first question is: "Should I go anywhere other than where I am right now?"

You will find that when you ask questions in this very simple direct manner that there will be a response deep within you, beyond your wishes and your preferences. And when you discover that Knowledge will not give you what you think you want, or will indicate that for you, you will begin to realize that there really is an independent wisdom within you that is not governed or manipulated by your desires, your preferences or your fears.

People are afraid to go to Knowledge because they are afraid it will discourage them from seeking what they think they want. But the very fact that Knowledge would do this is proof of its existence, that it is not a form of self-deception; it is not a creation of your imagination. There is really something independent within you that is beyond your own self-manipulation or the manipulation of others.

People call this intuition, but really that is not adequate to describe what Knowledge is really capable of. Fleeting experiences of intuition do not really tell you about the power of Knowledge within yourself; nor does it instruct you how you must learn to follow Knowledge and to be with Knowledge and to become simple and honest yourself in seeking this greater counsel.

Here it is not only Knowledge within yourself that is important; it is Knowledge within others. Certain people will tell you things that ring true. They resonate with you at a deeper level. Perhaps it is only part of what they have to communicate, but you listen for this in other people. You listen for the evidence of wisdom.

But you cannot be judgmental and condemning, or you will not be in a position to discern this in others. And if people do not represent

PREPARING FOR THE GREAT WAVES OF CHANGE

any degree of wisdom, you must not condemn them. For this is a world where Knowledge is rare, where people act out of habit and social convention, where people are manipulated and governed by forces in the mental environment.

Do not expect, then, to find Knowledge everywhere you look, for it will be rare. That is why people make foolish decisions and choose avenues of life that are self-destructive and that lead them nowhere. This is why people are unaware and unprepared for the Great Waves of change.

Even though the evidence is mounting all over the world, people still remain self-preoccupied, as if they are trying to insulate themselves from this growing reality in the world. They are afraid to know because they do not feel they have any competence in dealing with this situation. They do not want to know or be aware because it makes them uneasy; it produces a feeling of anxiety and insecurity, helplessness and hopelessness.

It is because people are not grounded in Knowledge that they are so easily frightened and that they are unwilling or unable to face the realities and the eventualities of their own lives.

People do not think ahead. They do not plan ahead. They either try to lose themselves living in the moment, or they believe that the future will be like the past. So they plan their future along these lines, amongst these assumptions. And how easily people will be overwhelmed and upset when they find out that life is really going somewhere else. They will not be in a position to see what is coming over the horizon or to discern it correctly or to respond wisely.

This is why Knowledge is the most important thing. You cannot stockpile food for the rest of your life. You cannot stockpile anything

LIVING IN A TIME OF UNCERTAINTY AND INSTABILITY

for the rest of your life. You cannot make a plan that will be effective in all eventualities. You cannot run away to some distant location, for people who are isolated will be very vulnerable in the face of the Great Waves of change. You cannot rely upon government or the economy to solve this problem for you.

You may even think that God will take care of you, but God has already put Knowledge within you to guide you. God is running the entire universe and a universe of universes. God is not preoccupied with managing your affairs. God has given you a perfect guiding Intelligence, yet it is deeper within you. It is not at the surface of your mind where you live every day.

Look at the world objectively. Ask yourself: "What is coming?" Listen. Write down what you hear and ask again and again over the course of days and weeks to see what comes into your mind. Listen more deeply. Do not be persuaded by simple assumptions that everything will turn out fine or that government or the economy or human ingenuity alone will take care of everything. Listen more deeply. Listen to your feelings of anxiety. Pay attention to your deeper experience.

Do not listen to the same old assumptions or reassurances that you tell yourself or that you have other people tell you. You must be willing to face discomfort here if you are to recognize the presence of Knowledge and to call upon Knowledge as your primary need. If you cannot face uncertainty, then you will certainly not be in a position to deal with reality.

This represents a deeper evaluation within yourself. It is not something you take on in conversation with other people. It is something you have to consider deeply within yourself, not searching for answers but instead being with the questions, being with the problem.

There is no simple solution to the Great Waves of change. You will have to work on these problems, live with these problems, be creative regarding them. This is bringing you back into living creatively in the moment, not living on assumptions, not living disassociated from yourself or from the world, in pursuit of hobbies or dreams or fantasies.

God has sent a New Message into the world to prepare humanity for the Great Waves of change—to alert humanity, to strengthen humanity, to teach The Way of Knowledge, to encourage unity and cooperation and to provide an understanding of your spirituality at the level of Knowledge, not at the level of belief or tradition or orthodoxy, but at the level of Knowledge as it exists in this moment.

The question is not whether you are religious or not; it is whether you are strong with Knowledge. The question is not whether you are a believer or not; it is whether you are strong with Knowledge. It is not a question about enlightenment or reaching some fantastically high state of consciousness; it is about being grounded in Knowledge.

Those who are seeking enlightenment may fail abruptly and tragically in the face of the Great Waves of change. Those who are firm believers in their religions may be overwhelmed by the Great Waves of change.

You have to deal with your environment—your physical environment, the natural world, your social environment, your economic environment. You cannot live in a fantasy within yourself, unaware and disassociated from this.

LIVING IN A TIME OF UNCERTAINTY AND INSTABILITY

Even the animals are always watching, always connected to their environment, always on the lookout for danger and for changing circumstances. Yet people walk around caught up in their own internal world, barely aware of what is going on around them, undiscerning about changing circumstances.

Do you see the problem here? People are lost in their dreams, their desires, their goals, their conflicts, their unforgiveness towards others, their regrets, their traumas. Here their intelligence is not serving them at all. They are unresponsive to the world. They do not see what is coming over the horizon. And they have not found the real source of their strength and certainty to even face such things.

When you look at the world, recognize you have a relationship with the world. You have a relationship with the future. You are here to serve the world. That is what has brought you into the world. That is your greater purpose for being here. You have a relationship with the future, for you were meant to be born at a time when the Great Waves of change will be coming.

It is no accident that you are here. Do not look at the Great Waves as just a giant inconvenience, a tragedy, a misfortune. You have come here to live in these times, to live facing these circumstances.

Knowledge within you knows this is true, and that is why it does not recoil from the Great Waves of change. Instead, it looks for the opportunity for contribution, and it seeks to position you in such a way that you can navigate these Great Waves of change, that you will not be overwhelmed or overtaken by them, that your life will not be destroyed by them, that your self-confidence and your self-determination will not collapse in the face of the Great Waves of change.

PREPARING FOR THE GREAT WAVES OF CHANGE

God knows what is coming to the world. Humanity is unaware and unprepared. God has put Knowledge within you, to prepare you, to give you the strength and the fortitude and the discernment and wisdom to navigate the difficult times ahead.

Therefore, get used to uncertainty. For you will be living in a world of increasing uncertainty, a world of increasing discord.

Do not condemn this world, for it is the world you have come to serve. Do not recoil from the Great Waves of change, for you must face them and allow Knowledge to guide you in order to discern what must be done and what must be done at this moment to prepare yourself.

Do not wait until the last moment. Do not rely upon consensus with other people. You must act now while you have time to prepare.

Is your career or form of work sustainable into the future? Does it provide essential services or goods for people? Can it survive a depression, economically? Do you live in such a place where you can get around without the use of an automobile? Do you have enough financial resources to weather economic difficulties? What is the status of your personal health, your physical health, your mental health? Are you in relationship with people who can become aware of the Great Waves of change and prepare for them courageously? Or are your associations with people who will just fall apart, become helpless and hopeless and dependent upon you?

You will have to take care of your children until they are adults, and perhaps your aged parents as well, but beyond this, you cannot have people hanging on you. For you have greater work to do in the world, and the Great Waves of change are the perfect environment for the discovery and the expression of this greater purpose in life.

LIVING IN A TIME OF UNCERTAINTY AND INSTABILITY

In that sense, they are perfect, but only in that sense because the Great Waves of change will be incredibly difficult and very tragic. Many people will perish. Many more will lose their wealth and their advantages. It will be a time of great tribulation.

Yet it could also be a time of great redemption for the human family. Nations and groups of people will have to cooperate to survive the Great Waves of change and to assure the distribution of resources. There will have to be much selfless activity and service to others.

It is only in facing a world in peril that humanity will really be able to establish the kind of cooperation that can build a new foundation for the future. So there is great promise here. But there is also great difficulty.

For yourself, it is important that you need as little as possible from the world, that your need for resources is minimal and that you owe others as little as possible, that you be as free from debt as possible. Simplify your life. Release yourself from financial obligations to the greatest degree you can.

Learn ways of traveling about without the use of an automobile. If you live in a rural setting, consider moving closer to areas where distribution centers for resources will be more available.

Strengthen your physical health and your emotional health. Set aside your hobbies and your interests or only maintain them minimally, for you will need all of your energy now to discern the Great Waves of change, to become educated about them and to prepare for them.

Be willing to face discomfort and anxiety within yourself, for you will be shaken by the truth. It will be shocking at first, and you will have times when you feel hopeless and helpless, where you have no confidence for yourself or for other people. But this will pass, and you will settle down as you adapt to these changing circumstances and to this growing reality in the world.

You will reach a place where you are just simply determined. You are clear and more objective. You have accepted that this is the reality you will have to face. And you will begin to reconsider the way you live, how you live, your obligations, your plans, your goals, everything.

People, particularly in wealthy nations, have lived in such a state of disassociation from their environment that they really will be helpless at the outset. They are so accustomed to other people taking care of things for them—their government, other institutions—that they feel quite incompetent in facing the Great Waves of change. They will say, "Someone should do something about this!" because that is how they have been living their lives, reliant upon other people taking care of everything for them. Yet now they will have to become more self-reliant, more competent, more present.

For poorer people this will be easier because they have had to face life continuously. Yet for the wealthy, it will be a very difficult transition, for they have been living in their own fantasies for so long that they have no real relationship with the natural world. Perhaps they have no relationship with the economic reality or the social or political reality. They have been dreaming their life away, and now they must wake up.

Everything will have to be re-evaluated in the face of the Great Waves of change. And here you will see that so much of what people are pursuing or are obsessed with will really have no value and meaning or will be greatly insignificant in the face of what is coming over the horizon.

LIVING IN A TIME OF UNCERTAINTY AND INSTABILITY

The question is: How will you function? How will you survive? How will you maintain your stability in the face of such a great convergence of changing forces? The emphasis here will not be on what will make you happy, what will please you, what will give you endless pleasure.

For your economies, it will not be an emphasis on growth. It will be an emphasis on stability. It will not be about making profits. It will be about sustaining one's own stability.

The ship that humanity lives on is slowly sinking, but people are having a party on the upper decks, or they are in their private chambers engaged in their own pursuits with other people or their own conflicts with other people. They do not realize the very foundation of their life is changing. The very circumstances of life are changing. For you are now facing a future that will be unlike the past.

So the first great threshold is to face the great threshold, to come to your senses, to awaken to a changing set of circumstances, to go through the shock and dismay and the great feelings of uncertainty or frustration this might produce.

Yet you must pass through this initial reaction if you are to be in a position to take action on your own behalf, in advance of changing circumstances. If you wait until the last moment, then you will have very few options, and anything you do will be really costly and difficult.

You must build your life on higher ground, higher ground within yourself, higher ground in terms of your circumstances in life.

Many people who engage in this deep evaluation will have to relocate, will have to consider changing their career. Sometimes they will have to leave the people who they are with, who cannot or will not respond.

Yet this will bring all of your life together. You do not have to worry about who you are and why you are in the world, for the world will be telling you what you must do.

Here you do not have to wander aimlessly and endlessly pondering your life and your wishes and your dreams, for life will be demanding things of you.

Here you do not waste your time in meaningless conversations with people or meaningless pursuits in the search of some kind of fleeting happiness or satisfaction, for life will be demanding things of you.

This will make you feel alive. It will make your life seem essential. It will force you to become realistic and observant and discerning. This is redemptive.

If you cannot come to yourself in times of luxury or affluence, you must come to yourself in times of difficulty and deprivation.

Knowledge is your guide. It is your compass. It is the center and the source of your life. It is intelligent. It is unafraid. It cannot be manipulated or persuaded. It is the center of gravity within yourself.

You see here that you will be able to rely on so few things on the outside. You will have to rely upon Knowledge within yourself and Knowledge within other people. You will have to reassess your life completely, and this will be redemptive if you can do this with wisdom and clarity.

LIVING IN A TIME OF UNCERTAINTY AND INSTABILITY

This warning is a gift of Love. It might arouse fear because you are unaware and unprepared and you do not yet know the source of your strength. But this is a gift of Love. God's New Message is a gift of profound Love for humanity.

Even if people respond with fear and anxiety or denial or repudiation, it is still a gift of Love. A parent may warn their children, and their children may scoff at it, but that does not mean the warning is not genuine and does not come from love.

The power of decision is within you. Knowledge lives within you. You have the power to prepare. You have the strength to become aware. You have the ability to overcome fear and helplessness and hopelessness within yourself.

The Great Waves will activate your life. They will focus your life. And they will call out of you your greater purpose and your greater gifts, which will not arise under normal circumstances.

This is the power of your time, the time for which you have come, the time that will change the fate and the destiny of humanity, the time that will alter the landscape of the world.

Do not worry about the outcome. Your focus is to give your contribution, to find it, to render it. But first you must bring your life into order and balance, and you must undertake this deep evaluation.

You must find your strength, and you must build your experience of this. You must shake yourself out of your dreams of self-fulfillment and your fantasies about your life and your pursuit of happiness and bring yourself back to the reality of your life and to your preparation for a future, a future that will be unlike the past.

As revealed to Marshall Vian Summers, October 8, 2008

CHAPTER 4

Preparing for the Future

At every moment you can live life to the fullest, but you are also here to prepare for the future. Preparing for the future, then, is an integral part of life. Every day you do this in many ways without even thinking about it. And yet consciously preparing for the future is very important and represents an aspect of intelligence.

To live only for the moment is not intelligent, for that is not what life asks of you. For you must be prepared for certain eventualities, and you must be prepared for certain mundane events, which you can even anticipate at this moment.

The future, however, can be overbearing. People often try to escape by believing that God will take care of them, or life will take care of them in the future, so they do not need to plan and to prepare for themselves. They think they do not need to make preparations or acquire provisions for themselves for the future, for God or life will take care of them. Those people try to escape into the present to avoid the burden of their anxiety about the future. But the future need not generate undue anxiety. Concern, yes, but not undue anxiety.

To be responsible to yourself and to others, you must accept that you have a fundamental duty to prepare for the future adequately. This has greater significance now, for you are preparing for a future that will be unlike the past. You are preparing for a world of declining resources, a world whose environment has been disrupted significantly, a world where there will be climate change and violent weather. A world where the risk of competition, conflict and war over the remaining resources will be extreme and very dangerous.

If your life up to this point has not been unduly difficult, then perhaps you will assume this will be the case in the future. But that would be a very unwise assumption.

Your task now is to bring wisdom to bear on what must be done today, and tomorrow and the days to follow. It is not a decision between love and fear. It is a decision between responsibility and irresponsibility. It is a decision between wisdom or anxiety.

For if you do not prepare for the future, you will have anxiety. You will feel that you are missing something. You will feel uncomfortable with yourself and your circumstances because you are not yet responding to what Knowledge, the deeper Intelligence that God has given you, is indicating that you must do and that you must attend to.

So there is no real escape from this responsibility. For if you try to escape, you will feel anxiety, and you cannot escape this anxiety, for it is telling you that you are being irresponsible, that you are not meeting the requirements of life at this time which is you live in the moment and you prepare for the future appropriately and accordingly. If you think the future is going to be like the past, well, whatever plans and provisioning you do will be greatly inadequate.

Today so many people are resting on assumptions that the future will be like the past. It is upon this that they make their financial decisions. It is upon this that they determine what they will do and how they will use their resources and what they will commit themselves to.

PREPARING FOR THE FUTURE

But you must know that the environment is changing. Circumstances are changing, even for the wealthy nations. You must be attentive now and become as objective as you possibly can.

When you begin to face this reality, there will be moments of anxiety and apprehension. You will feel perhaps inadequate and overwhelmed. This is understandable. Yet you must get beyond these initial feelings, for they are simply reactions. They reveal to you how unprepared you are even at this moment for the eventualities of life.

You must gain a sober approach. Roll up your sleeves and say, "Well, what must we do here?" And lay out a plan with many steps and begin to do the easy things first. Take care of those, and then take on more challenging projects.

At the outset, you may be completely confounded as to what to do. But this will change as you begin to take steps to correct or change or prepare for immediate things that you can recognize that are well within your comprehension now. That will take you to the next step.

This is not a process where you go out and spend a weekend and buy a lot of things. This is not something where you simply make some grand assumptions and set the whole thing aside for yourself so that you can go back to your previous preoccupations.

This is a process that has many steps. For you must prepare now for a future that will be unlike the past—a future that is far more unpredictable, a future that is far more changeable and a future that is far more dangerous and yet, a future that will provide some remarkable opportunities for you and will greatly assist you in developing your strengths, your competence, your skills, your awareness and your discernment.

PREPARING FOR THE GREAT WAVES OF CHANGE

If life is too easy, these things never become cultivated, and the whole focus on the path of enlightenment becomes skewed and incorrect. People seek to have greater peace and equanimity. They want to feel better and feel better and feel better, and the reason they do not feel better is in part because they are not preparing for the future and they are experiencing anxiety and discomfort.

For if you do not follow what Knowledge, the deeper Intelligence within you, is indicating, you will be uncomfortable, you will be apprehensive, you will be nervous because you are not following what the deeper Intelligence within you is indicating. You are not paying attention. You are giving your attention to the wrong things, or you are pursuing things for the wrong reasons. Or you are neglecting, in any case, your primary responsibilities.

People think there is something wrong with them when they are feeling uncomfortable, and they try to get comfortable, so they try to run away from the very thing that is trying to communicate to them. The very advice, the very message, the realization that is trying to reach them, they want to now run away from this. They want to be comfortable. Such foolishness. And so they compound their problem. Now they are even more uncomfortable.

There is no escaping your responsibilities. God has given you this greater Intelligence, the Intelligence of Knowledge, to guide and protect you. You have a fundamental responsibility to build a connection to this Knowledge, to connect your thinking mind—your personal mind, your worldly mind—with the deeper Divine Mind of Knowledge within yourself, which is trying to guide and prepare you, to take you to the right places, the right people, the right circumstances and to free you from the burden of all the things you have added on to your life which are non-essential or which are harmful for you.

PREPARING FOR THE FUTURE

So do not try to be comfortable. Try to be responsible. If you are really uncomfortable, there must be a reason for it. Ask yourself, "Why am I feeling so uncomfortable? Is there something I need to see, know or do?" Keep asking yourself these questions until you make a connection within your own experience.

The longer you are in denial, the longer you remain obsessed with non-essential things, the longer you are involving yourself in relationships with people, places and objects that are inappropriate for you or which are weakening you and taking your attention away, the more you will feel out of sync with yourself. You will feel disconnected from your own deeper experience.

Indeed, this is the condition of most people in the world today. They are uncomfortable. If they are not in terrible circumstances already, they are uncomfortable.

Even the wealthiest people—very uncomfortable. Now they want to buy more things, have more stimulating experiences, go see more movies and have more exciting adventures or more romance. They are simply delaying the recognition that there are things they need to see and know and do that they are neglecting. If this persists, it becomes very chronic and very disabling.

Your first responsibility in life is to build a connection to Knowledge, to take the Steps to Knowledge, for this is how God is going to help you. You can pray to God for many things. You can pray for success. You can pray for health. You can pray to avoid disaster. You can pray for your friends and your family and loved ones. But if you want to allow God to guide you, you must build a bridge from your worldly mind where you live to the deeper Mind of Knowledge within yourself. For it is Knowledge and only Knowledge that can guide and protect you in the difficult and uncertain times ahead.

Fervently believing in God, fervently believing in the teachings of the savior or the prophet or the enlightened one will not prepare you for the future and will not provide you the wisdom of what you must see, know and do now to live your life responsibly.

These fervent beliefs will not bring you closer to God unless you can build a bridge to Knowledge, for it is through Knowledge that God's Will can speak to you in your circumstances. It will not be God speaking like a person speaking to you. It will be the thoughts, the images, the urges, the restraint, the encouragement—all coming from a deeper, more mysterious Intelligence somewhere within you, somewhere beyond the grasp and control of your intellect, your ideas and your beliefs.

If you want to be close to God, you must build a connection to the wisdom that God has placed within you. If you want to allow God to assist you, you must build a connection to Knowledge and take the Steps to Knowledge and build this foundation in Knowledge.

For only Knowledge knows what is coming. Only Knowledge can tell you what to do at the difficult moments of decision. Only Knowledge will take you to the right people, the right places and the right situations and circumstances. Only Knowledge will pull you out of disaster and prevent you from leading yourself to disaster.

God does not need your belief. God does not need your worship. God does not need to be celebrated. God needs to be followed.

You follow God most profoundly, most purely, not by believing in the edicts of religion necessarily, but by building a connection to Knowledge. If religion is functioning correctly, it is helping you to build a bridge to Knowledge, to your own conscience and to

PREPARING FOR THE FUTURE

live responsibly, ethically and beneficially, both for yourself and for others.

When religion does not do this, it is not serving its primary purpose, the purpose that was established by God. But in so many cases, religious leaders and institutions have lost this fundamental understanding, and now they use religion for other purposes—for political purposes, for economic purposes, to corral people and to direct their thoughts and behaviors for the interests of the state or the interests of a church or the mosque or the temple. This is not religion's real role and task. It is an aberration. It is a misuse. And in many cases, it is an abomination.

At this moment you do not know what to do completely to prepare for the future because you do not know what kind of future you are preparing for. And even if you could find that out and have greater clarity here to know what is coming over the horizon of your life, you do not know all the steps that you will have to take to prepare for this.

Therefore, you are going to have to learn to be attentive and patient and responsive. There are things you must do today and tomorrow and in the days to come. Right now you cannot see all of it. But these things will reveal themselves to you if you can proceed accordingly, building your connection to Knowledge and preparing your life for the Great Waves of change that are coming to the world.

For humanity has two great challenges. It has to face the results of centuries of environmental decay and destruction, the results of its own misuse of the world, the results of its lack of wisdom in the past. The results of so many of these things will be coming due. It will be a great threat to the well-being of people everywhere, even in the wealthy nations.

PREPARING FOR THE GREAT WAVES OF CHANGE

The second great challenge is humanity's contact with intelligent life from the universe, from the Greater Community of life in which you live. A dangerous Intervention is already occurring in the world and has been underway for several decades.

Together these represent the greatest challenge to humanity's well-being, freedom and sovereignty in this world than has ever occurred before. An unprecedented set of circumstances, when taken together, can seem truly overwhelming and daunting, overwhelming and daunting to your mind and intellect, but not to Knowledge.

For you have come into the world to meet these challenges. You have come into the world to serve the world and to assist humanity and all life here according to the specific role you are meant to take and the specific people that you must find and build a true connection with.

It takes a Greater Plan here, not a plan that you make, not a plan that you invent, not a plan that you read about in the newspaper or the magazine.

At any given moment in following Knowledge, certain questions will be answered and certain questions will not be answered. If you are to take such a journey to gain such wisdom, to learn of a Greater Plan of which you are a part, you must have an openness of mind. You must avoid making premature conclusions and living on assumptions. You must have the courage to face the uncertainties of life and the great uncertainties that the future will create for you and for everyone.

Here you must accept that the things that you believe will provide for you and take care of you may not be there for you in the future, or for your children and your loved ones.

PREPARING FOR THE FUTURE

The supermarkets may not be full of food. The government may not be able to help. There may not be medical care available to you at all times and all circumstances. You may not have a job. There may not be a job. The environment will be changing.

How people respond to this remains to be seen, but certainly there will be social disorder. Certainly, there will be great political and economic instability. Certainly, many people will not respond wisely or appropriately to these circumstances.

Many people's circumstances, their personal circumstances, will change dramatically, even very suddenly. It will be great travail for them, for they are unprepared. They did not see what was coming and they did not prepare.

You will see this all around you, and it will be very disheartening and, in many cases, very tragic. That is why preparing for the future is of such great importance for you because you must be in a position, not only to navigate the difficult times ahead, but also to support and assist others who will be facing great difficulties.

Therefore, do not be dismayed that you perhaps are the only one you know who is taking these preparations, who is looking into the future, who is seeing what the New Message from God is revealing about what is coming for humanity. You may be the only one who is looking and reading and studying the signs of the world, the signs of nature, the cues and messages that you are receiving, that you must receive and recognize and face.

Many people will not look. They will not prepare. They will live on their assumptions. They do not want to be troubled. They do not want to put in the extra effort. They do not want to deal with anxiety and uncertainty. They do not want to have to make difficult decisions or give up things that they think they want or need for themselves.

Right now you cannot help them. You must build your own foundation, a foundation in Knowledge and a foundation in the Four Pillars of your life—the Pillar of Relationships, the Pillar of Health, the Pillar of Work and the Pillar of Spiritual Development.

You must build your own life raft if you are to be in a position to assist others. You do not want to be a victim of the future. You want to be a beneficiary of the future and a contributor in the future. That means you have to prepare now, for the hour is late, and the Great Waves of change are coming.

Already, you can feel the impacts. Already, if you look around the world objectively, without insisting on your beliefs or ideology, if you look objectively, you can begin to see the evidence that great change is just beginning.

This will perhaps be disconcerting to your mind, but to your heart it will be a confirmation, for this is why you have come. You did not come all the way into this world to give up your Ancient Home and your Spiritual Family to enter here to simply acquire pleasantries for yourself, simply to enjoy life's sweets and life's many little pleasures. You came for a greater purpose. And in your heart you know this to be true.

You do not want to allow yourself to become decadent and corrupted by these pleasures, even the ones that seem very simple and harmless. They are fine in their place, but they cannot be the focus of your life. If they are, it means you do not know your life. You are not connected to Knowledge. You do not know your own heart. You are not properly engaged in the world at this time.

PREPARING FOR THE FUTURE

So you must find the courage within yourself to face the Great Waves of change and to keep looking to allow your understanding to grow and to become more complete. You must begin to take the steps that you will have to take to fortify your life, to alter your life, to reinforce your life, to redirect your life and to prepare for what you will see is coming over the horizon.

Here you cannot simply pack up a lot of food and go hide somewhere. That will not work. You cannot simply move to the country and think you will be safe, for you will not be safe there. The Great Waves are too great.

There is no place to hide. All the reassurances you give yourself will be inadequate to deal with the circumstances that are coming to you. You are going to have to become more sober and more serious about this matter, more careful and more patient.

As We have said, you do not know all the things you will have to do, but you must begin the process—begin to take the Steps to Knowledge. Begin to assess every activity you have in your life and ask yourself: "Is this activity appropriate for me and will I be able to maintain it in the near future?"

Go around your house and ask yourself about every possession that you have: "Does this possession really serve me? Will it serve me in the future?" Begin to sort out, for you cannot be given a new realization if your life is full of relationships with people, places and things that are not right for you, or that will not help you.

Every relationship you have should give you the strength to do this. Every possession you have should either be useful or inspiring to you. Every commitment and obligation must help you to prepare and to gain a foundation in Knowledge.

PREPARING FOR THE GREAT WAVES OF CHANGE

For there are no neutral relationships in life. All your relationships with people, with where you live, with your possessions, with your activities, are either strengthening you or weakening you in taking the Steps to Knowledge and in building a real foundation for your life.

Here you must learn to become inner directed rather than outer directed. Here you will find out as you proceed that even the experts really do not know what they are doing—those people who seem so self-assured, who have such strong views or vaulted opinions—they are only living on assumptions. These assumptions are very doubtful.

Here you will see that you cannot simply follow someone, believing that they know the way. You can only follow someone who is strong with Knowledge, but if you are not strong with Knowledge yourself, how will you know who they are? How will you know how to be with them? They will not lead you around as if you were a little puppy, a little pet.

You yourself must become strong, stronger than you are today—not fearful, not full of trepidation, but clear, objective, open and observant. All the energy you spend today condemning people and governments—all that energy now has to be saved to prepare you for the future. All the time you waste on your hobbies and your meaningless conversations with people—all that energy is going to be needed for you now. You cannot be wasting it in all directions.

Every day is important. Every month is important. Every year is very important. You have your wonderful times and your pleasant moments, and as you proceed, you have less and less anxiety because your life is becoming [more] in keeping with Knowledge within yourself. You actually feel freer and more confident, whereas before your life was governed by fear and avoidance.

PREPARING FOR THE FUTURE

When you are moving with life, you gain a greater confidence in life and a greater confidence in yourself that you will know what to see and do. But you must be moving with life. You must be building and learning to use your time and your vital energy meaningfully and appropriately.

You cannot sit on the side of the road and have someone else build for you a foundation for the future. Many people will try that, of course, but that will not lead to success, and for many people it will be disastrous.

You must learn to be open. You cannot control this process, but you must control your thoughts and your mind and your actions. That is where your self-control really will make the difference. You must learn to be able to restrain yourself and hold yourself back from giving your life away in premature relationships or in outrage against the world or in following people who really do not know where they are going or what they are doing.

Do not assume that others know. You must know. Not all the facts. Not everything about the future. But you must sense what is coming, and you must realize what you have to do to respond to that, what is within your power to do each day. Then, and only then, will you begin to feel that Knowledge is a presence and a power in your life, and not simply some remote possibility.

Here it is necessary to get people moving. If people are asleep by the side of the road, or they are standing around not knowing what to do, you have to get them moving. You have to get them involved.

We are talking about you. Your life has to become uniform and strong. You have to see where you are losing energy to people, to situations, to obligations, to activities, to jobs that are really not right for you. It will take a lot of strength and a lot of effort to make the necessary adjustments to your life.

You must have this strength and this energy. In gaining this, you will see how you are wasting your strength and your energy, how you are giving away your life to others, how you follow needlessly and heedlessly what other people want you to do or tell you you must do. That will not help you now. In fact, it never helped you at all.

You will need to learn to live using much less energy. You need to live near where you work. You will need to be able to get around without the automobile to a far greater degree than at present.

You must be employed in work that has a future in a declining world, providing fundamental services to people—goods and services that people will always need, not simply what they can afford with their spare time or their spare money.

You must build some real skills. You will probably need to learn how to grow at least part of your own food and learn some basic skills that have been lost in your modern cultures.

Your life will have to become much simpler than it is today, much more efficient than it is today. You will have to learn how to use your time and your financial resources much more carefully, and to find pleasure and joy in simpler things.

You will need to own far less, for you will not have room for things in the future. You will need to have some savings. You will need to have financial resources available to you. And you will need to have such strength that you can assist others who have genuine needs, the very young and the very old in particular.

PREPARING FOR THE FUTURE

This is what will redeem you to a true life, to an authentic life, not a life of obsessions and addictions, a life of personal avoidance and irresponsibility, but a real life with real relationships, meaningful work and deep, simple pleasures.

Humanity has used up much of its endowment in the world. Now it must face the consequences. It is also facing competition from the Greater Community in the form of Intervention from groups who are here to take advantage of a weak and divided humanity for their own purposes. You must learn about these things, as bravely and as objectively as you can.

Regarding the Greater Community, only God's New Revelation can show that to you, for humanity is ignorant and vulnerable in the Greater Community. You will need a New Revelation to understand what you are facing and what is occurring here now. Such an understanding you cannot find in your textbooks, your colleges, your universities or your popular magazines. Even in esoteric writings, the truth of this is not clearly revealed.

This may seem overwhelming at first, but it is what will save you from an insipid and mundane life. It is what will free you from bondage and obligation to others that can only weaken you and waste your time and your energy.

Allow the challenges of life to clarify your life, to unify your life and to strengthen your life. Do not worry about what will have to be given up. Things fall away along the way.

Already there are things you are doing that you are ambivalent about because they are not really meaningful. You are only involved to please some expectation of yourself or someone else's expectation of you. Preparing for the future will free you from the ravages of the past and from the dissolution of your life.

You want God to save you? You want God to redeem you? You want God to give you strength, purpose, meaning and integrity? Then this is it.

Knowledge will carry you forth, and anything that is not of Knowledge will fall away or become far less important to you as you proceed. This is how your life is brought back to its Source. This is how your activities are given clarity, uniformity and direction.

You must face the future or it will overtake you and destroy your life. Do not think that because you live in a certain country or you believe in a certain political system or you adhere to a certain religion that you will be exempt from this. There are no exemptions.

Before coming into this world, you knew you were entering a time of great change and difficulty. You came with this understanding. That understanding is still within you, kept for you by Knowledge, the great endowment God has given you—this deeper Mind. It has the memory of your Ancient Home and the memory of those who sent you into the world.

But your intellect cannot find this memory. Your worldly mind cannot grasp it and claim it. It will reveal itself as you proceed, as you take the Steps to Knowledge within yourself and as you begin to focus your life on the outside.

God's New Message has been given to alert you and to prepare you and to bless you. It is here to alert you to the realities of life both now and in the future—realities that most people cannot see or will not see. It is here to prepare you by teaching you what spirituality

PREPARING FOR THE FUTURE

means at the level of Knowledge, a Teaching that has never been given to all of humanity before. And it is here to bless you, to give you the strength, the protection and the guidance that only Knowledge can provide.

For Knowledge is of God. It is not a human invention. It is not something that was made up by some institution or a group of people. It is the foundation of your life. It is your guiding light.

It is Knowledge that will give you the courage and the commitment to follow what you most deeply know and to do what you see that you must do. This is God's blessing for you. It has always been with you. And now you need it as never before.

As revealed to Marshall Vian Summers, September 5, 2007

CHAPTER 5

Adapting to a Changing World

In considering the Great Waves of change that are coming to the world and humanity's encounter with intelligent life in the universe—the two great events of this era of human evolution—it is necessary to understand what one must do to adapt to these changing circumstances. For even if humanity musters a great effort to counteract the impact of a deteriorating environment; of a changing climate; of diminishing food, water and energy resources; and even if humanity were to gain a greater unity and cooperation in facing the reality that your world is being penetrated by forces from beyond, from the universe in which you live, you will still have to adapt to changing circumstances.

This adaptation then must be accepted, for if you think that technology, or some kind of magic, or government policy will be able to prevent this requirement for adaptation, you will be making a very serious mistake and will be denying yourselves the time necessary to prepare for a great change in your circumstances.

Indeed, what are you preparing for? You are preparing for a world that will experience grave shortages of food and water in certain areas, and overall will have to face diminishing resources concerning energy, and even very basic materials that you rely upon and that people everywhere rely upon. This is the consequence of humanity living beyond its means for so very long. It is as if the bill has come due after one has borrowed so heavily from your natural inheritance.

Many people today, of course, sense that deprivation will be coming and that things will have to change very dramatically. But often they do not see the scope of this, thinking that it can be remediated, thinking that it can be moderated by humanity's efforts now. And, of course, these efforts are vital and necessary and should not be undermined for any reason.

But even so, great adaptation will be required. You will have to learn to live on far less. Unless you are impoverished and cannot feed yourself or house yourself adequately at this moment, you will have to use your resources much more carefully.

Of course, this will change your economy. It will change the appearance of life around you, as people now seek to fortify their own position with the very fundamentals of life and to build relationships not based upon hobbies or interests or consumer activities, but based upon deeper alliances that will be necessary for security, for cooperation and for the sharing and use of these resources.

The impact of this will be very great, and the adaptation will be difficult. Do not underestimate this. It will be greater than you now think. Yet has this not been recognizable, even from an earlier time? Could you not see that humanity would use up so much of the vital resources of the world, setting the world into a permanent state of decline, thinking that technology or innovation alone would be able to mitigate the results of this?

Have you not felt, perhaps for a long time, that there would be great consequences to how nations, groups and people use the world—diminishing its resources, destroying its wildlife and using its vital

ADAPTING TO A CHANGING WORLD

energy resources without any concern for the future, without any focus on conservation? Have you not seen this and felt this in moments of clarity or introspection?

If you can account for these past experiences, you will see that what the New Message is presenting is not so very new, but has actually been part of your experience for a very long time. It has been with you, though you have not adequately been with it.

The world has been speaking to you. The signs of the world have been speaking to you, but you have missed so very much, concerned only with your interests and your problems, your conflicts and your concerns, missing the cues of the world, not recognizing the movement of things.

For humanity has already turned a corner some time ago, and now it will have to face the consequences of diminishing resources in so many areas. The more you are connected to the world in this regard, the more vulnerable you will be. Even the very simplest things that you use—resources from around the world, resources that you expect and have depended upon, never thinking where they come from, how they are produced or the cost to nature and the environment for their production—now these things will become an ever-greater concern and produce ever-greater consequences for people everywhere.

What will this look like for you, who now must concern yourself with preparing for a future that will be so unlike the past? Think first that you must reduce your consumption of resources by at least fifty percent, particularly if you live in a wealthy nation or an affluent lifestyle. If you are very rich, you will have to reduce your consumption even more.

PREPARING FOR THE GREAT WAVES OF CHANGE

Your willingness to do this represents your integrity and your concern for the world. For if you insist upon an affluent, luxurious lifestyle, you will be feeding the engine of war. You will be requiring your government to gain access to resources wherever it can, at whatever cost—often beyond any boundaries you might set ethically or morally for yourself.

You are adapting to a world where it will be more difficult to live. And you will require greater innovation in your technology and a greater cooperation between your nations. But even here, you will not be able to offset completely the great changes that will come to the way you live and to your priorities.

There will be regions of the world that will become uninhabitable, and millions of people will have to flee them. They will become too arid. Their ability to sustain people with food and water will diminish to a point where people will have to leave and have to escape.

Where shall they go, these millions of people? They will need to find new homes, seeking access to the more temperate regions of the world and to the more affluent countries of the world. Will you accept their presence, or will you struggle against them to protect your lifestyle, to protect your own priorities? And what will happen when nations cannot sustain themselves economically because of the loss of resources, when the demands of their people far exceed the supply of what they can provide? How will you regard this?

How will you respond when the cost of your food will take such a great proportion of your income? And will you have work within industries that will have no future, that society cannot afford? What will be your position? What will be your preparation? Will you pay attention to the signs of the world and begin to consider your life

ADAPTING TO A CHANGING WORLD

seriously, with commitment and compassion? These are all important questions that you must answer for yourself.

If you pay attention, the world will tell you what is coming. Knowledge within you, the deeper Intelligence that God has given you, will indicate the steps that you must follow—the beginning steps and all the steps that proceed thereon.

Perhaps this will require only minor adjustments, but you should consider that the change that will be required of you will be very substantial. You may not be able to live where you live. Your employment may not be viable in the future. The cost and expense of living will be very significant. And the needs of the very poor will be ever greater than they are today. That means that you must be prepared not only to provide for yourself, but to provide assistance for others—whether they be in other countries, or whether they be in your own neighborhood.

This adaptation is critical for humanity's well-being, survival and potential for the future. For there must be a very great change in how people live and how they regard their relationship with the world and with one another. As things stand today, this change will be brought about by a few courageous people, but the majority of people will continue to plunder the world and to demand and expect ever-greater things from their governments and their religion, from one another and even from God.

Only a few will be visionary enough and courageous enough to really prepare for the future, and their preparation will make an immense difference in the kind of future that they have. They must be willing to function without consensus, doing things that other people are not doing, making decisions that others would not consider to be important—reducing their expenditures, reducing their activities to only what is really meaningful and essential.

Here entire industries will disappear, as people no longer have the resources to indulge in them. Hobbies, travel, art, luxuries, collecting things. These industries and all the services associated with them could largely disappear.

Begin to think about this. Think about the plight of the elderly. Who will take care of them? Think about the plight of children. Who will take care of them? For if you do not prepare for the future, it will overtake you. If it does, you will not be in a position to provide much of anything. Instead, you will yourself require and need great assistance. Who will provide this great assistance? This is a very serious matter. You must have great courage to face it. And face it you must.

This is why God has sent a New Message into the world, to prepare you for the great change that is coming. What it speaks of is beyond current human conversation, except in very, very select circles. What it reveals is beyond the scope of human awareness. What it provides is greater than what humanity can provide for itself. Yet the New Message is here to encourage a greater honesty, a greater compassion and a greater awareness of what is happening in the world and what you must do now to begin to reorient your life and prepare your life.

Ultimately, this must come from Knowledge within you. For God has placed Knowledge within you to guide you and protect you in times such as these and into a future that you can barely even consider even at this moment.

It is to build your connection to Knowledge that is the greatest gift the New Message can give you, for you will have to rely upon it so

ADAPTING TO A CHANGING WORLD

very greatly. And it will reveal to you what your intellect could never understand. It is how God will speak to you—guiding you, counseling you and reinforcing you as you proceed forward.

Dispel your fantasies about transformation. Dispel your fantasies that you are entering an age of abundance. Dispel your fantasies that you could alter this with your thoughts, or with your affirmations, or with your proclamations. You have a fundamental responsibility to the world, and the world will determine the degree to which you can live here. You cannot override this.

For some people, this will require a great rethinking of their ideas and beliefs and position. Others will rail against God for letting them down, for not providing for them, and there will be a great loss of faith. Others will strike out against their neighbors, their governments or the governments of other worlds, thinking it is all a matter of politics and economics, failing to see that they have violated their fundamental relationship with the world and with nature itself.

God has put you into this world, but God has also set in motion the forces of nature. They are governing forces, and they are restraining forces. You can only overcome them to a certain degree. Beyond that, you must pay attention to them, honor them and understand how they function.

If humanity destroys its fundamental energy resources, what shall God do for you? What God will do for you is what God has already done for you, and that is to place Knowledge within you. Our recommendations are only for the outset, to give you time to see, to know and to prepare. But it will be Knowledge itself that will enable you to navigate the difficult and uncertain times ahead.

PREPARING FOR THE GREAT WAVES OF CHANGE

This is not about being positive or negative, fearful or loving. Set aside these dichotomies, for they are foolish. It is whether you can see and feel the movement of things, and whether you can respond as objectively as possible, without the blinding influence of hope or fear, but to see clearly. That is the essence of the matter.

You must be willing to see and feel things that others do not see and feel, functioning without agreement and consensus if you are to be strong with Knowledge. If you wait for everyone else, you shall share everyone else's fate and predicament.

No one is going to come and rescue you. There is no returning to a golden age in the past. There is no magic formula or secret technology or extraterrestrial gift that is going to take this challenge away from you. Give up demanding solutions and face reality.

You are living at the end of the age of indulgence and now must enter the age of human unity and cooperation. The age of indulgence is coming to an end, and for many it will be overwhelming. Do not be overwhelmed. For many it will be a disaster. Do not allow yourself to go through that disaster. For many it will be such a profound disappointment that they will not know what to do. Do not be amongst their number.

Search your religious traditions. Search your worldly wisdom. Search the history of humanity. Search the New Message from God. Search for the signs of the world. Search for the signs of Knowledge within you that are even at this moment speaking to you and urging you to move in a certain direction.

Take the easy path and you will want to forget, thinking it is all foolishness or that it comes to nothing. Take the difficult path and you will have to face great unanswerable questions and a great deal

ADAPTING TO A CHANGING WORLD

of uncertainty with only Knowledge within you and Knowledge within others to lead you forward.

This is your greatest promise. This is your greatest hope. Do not appeal for the governments to take these problems away, for to a certain extent, they will not be able to. Do not require that God remove the results of centuries of humanity's abuse of the world, for you and your children and their children must face the consequences of this.

You are living at the end of the age of indulgence. It is beginning now the age that will require human unity and cooperation and immense human courage and ingenuity—not the courage and ingenuity of a few saintly people, or a few gifted people, but of you and your neighbors.

The great temptation will be that humanity will fall into competition, conflict and war over the remaining resources. People will fight with each other to get what they want, what they need. This will happen at a local level, at a regional level, at a national level and in the world at large. This great temptation to fight and struggle, to overcome others to acquire what you want or need—these tendencies are already being activated. The fire of conflict and war is already being stoked in many places. The fear of deprivation is already overcoming many people, some of them very wealthy.

There must be a great choice in how humanity will face both the Great Waves of change and Intervention from the universe. If you struggle and fight, your chances of success will diminish accordingly. If you unite and join your resources, your possibility for success is enhanced accordingly.

PREPARING FOR THE GREAT WAVES OF CHANGE

That is a new platform for peace—not a platform based upon ethics or morality alone, but on absolute necessity. For no nation will be supreme if other nations collapse. No nation will be immune if the society, the economy and the social structure of other nations collapse. There is no immunity here with wealth and privilege. In fact, the wealthy and the privileged will have so much more to lose, and will feel so much more threatened by the thousand faces of the Great Waves of change and the thousand possibilities of conflict and collapse. They will even face the hostility from the poorer peoples, who will look upon them with hatred and vengeance.

Clearly there must be a New Message from God to help offset these deep-set tendencies and to minimize the effects of competition, conflict and war. It must stir within people a greater compassion and a greater commitment to secure security not just for their nation or group but for humanity itself.

For humanity as a whole is entering the most dangerous period of its entire history, a period in which its entire future could be determined within the next three or four decades—a set of circumstances where not only the well-being and the security of humanity is at stake, but your freedom within a Greater Community of intelligent life. Do not think this will not affect your life, and even very profoundly.

Therefore, your ability to see, to know and to prepare will lessen the difficulty, will lessen the stress of this, will give you a more solid ground to stand upon and will put you in a position to assist others. For there will have to be great human contribution in the future, greater than has ever been required before.

This has a redeeming quality to it. For everyone in the world was sent here to serve a world in need, to serve the world in these very circumstances that are being revealed here. Therefore, no matter how

ADAPTING TO A CHANGING WORLD

difficult things can become, no matter how challenging circumstances become, it has a redeeming quality in bringing people into greater service to one another and to the world.

This is the great potential of your time, the great promise and advantage of your time, but it can only be realized if you have a stronger foundation within yourself that can keep you from building fantasies or responding from fear, terror, anger or revenge. God has given you this foundation, and though it is little known to you at this moment, its value and importance to your life will become ever greater and eventually will be the focus of your life.

If everything were wonderful and people were secure and everything were assured, the need for Knowledge would not be great and only the very wise or the very dissatisfied would seek it for greater revelation and greater fulfillment. But entering a time of profound and prolonged difficulty is actually a very good environment and very stimulating for Knowledge.

For you must wake up now. You must become serious about your life. You must pay attention to your circumstances. You must learn about what you use from the world and how you are going to deal with profound and unexpected change in the future.

This is very redeeming for people and can ultimately make humanity a much stronger and more united race than it is today. For your affluence has been more of a curse than a benefit for many people— leading them to dissolution, leading them to corruption, leading them away from the power and presence of Knowledge within themselves, making them listless and unresponsive to the world.

PREPARING FOR THE GREAT WAVES OF CHANGE

The age of indulgence is coming to an end. Its dangers are immense. Its opportunities are immense. Its challenge will be overwhelming. Its opportunity for contribution will be profound.

You cannot hold yourself back here. You cannot remain neutral, disassociated from all these things. And the degree to which you can recognize this now, accept the shock of this now, will have a great bearing on whether you can survive in the future and build a strong foundation for your life and fulfill yourself through service to others—to become a person of integrity, a person who is deeply responsible, a person who has gained a greater worldly wisdom, a person who can feel the grace and the power of Knowledge within themselves in times of peace and in times of great difficulty.

You have a great chance, a great opportunity here to emerge from these difficult circumstances a renewed person—a person of great strength and vision, a person who is capable of facing uncertainty, a person who can view conflict without hatred and anger, a person who can see human need and recognize ways that it can be met, a person who is not fooled by all the things that fool people and make them weak and unsuspecting and easily manipulated by others.

You have this great opportunity, and now the world will support this opportunity by requiring great things from you. Do not, then, feel sorry for yourself that you must face such a great challenge in life, for indeed it is a gift. It holds the promise of your redemption.

For you are not redeemed in the world by believing in God, or by worshipping God. You are redeemed in the world by fulfilling what you came here to do. And what you came here to do is tied directly to the condition, circumstances and future of the world. No matter what the nature of your contribution, even if you are only to serve one other person, this will still be the case. But it takes a change of heart,

ADAPTING TO A CHANGING WORLD

a shift within you to see the great possibilities for you and the great possibilities for the entire human family.

For I tell you, weak and divided, you will not remain free in the universe. Other groups, intervening groups, will gain access to your leaders and to the sources of power in this world. If you are weak, indulgent, divided and in conflict with yourselves, you will be fundamentally weak and vulnerable in the universe.

So the great changes that are coming to the world now, and the Great Waves of change, have the opportunity to re-establish humanity as a powerful and united race of freedom-loving peoples. Indeed, it is only the Great Waves of change that really at this point hold the promise of giving you this possibility. Without this, humanity would simply decline—corrupt, conflicted and indulgent. It would just diminish, until some other force in the universe came along to claim authority here.

So while you are dreaming about your life, dreaming about fulfillment, dreaming about the things you want and fearing the things you do not want, there are great forces at work in the world—moving the world, changing the circumstances of life. Ignore these at your own peril.

Recognize these. Face these. Do not demand solutions, for you must work with the problems. You must gain access to others to help you. You must become strong, stronger than you are today, wiser than you are today, more sober about your life—which means you are not governed by hope and fear, but can see clearly, objectively, with courage.

This is the great Revelation of your time. It is towards this that you must give your attention now. You have time, but not a lot of time.

PREPARING FOR THE GREAT WAVES OF CHANGE

You have an opportunity, but not an endless opportunity. You have real promise, but not endless promise.

Receive God's Message and Revelation and God's warning, blessing and preparation, for it has come to the world at this time. It calls upon the great well of human wisdom and compassion that has been built over the centuries despite humanity's conflicts and abuse of the world. You have everything you need to be successful. Your greatest adversaries are yourselves.

Therefore you must choose. This choice is not simply an idea but a pathway that you follow, your life demonstrating which way you have chosen—what you have chosen for yourself and for the world and for the future of the human family here. That is your statement.

Make no verbal proclamations. But look to your life and look to what is coming over the horizon—without hope and without fear, but with the clarity of Knowledge.

As revealed to Marshall Vian Summers, May 22, 2008

CHAPTER 6

ESCAPING FEAR, CONFUSION AND HOPELESSNESS

If one is really honest with oneself, one must realize that the world is changing and that Great Waves of change are coming to the world. This is more of a feeling within oneself: a sense of apprehension, a sense of anxiety, a feeling that one must prepare for something, even if one does not know what it is specifically.

For there are Great Waves of change coming to the world—great environmental changes, the diminishing of resources, growing economic instability and the ever-increasing risk of competition, conflict and war over the remaining resources.

Perhaps you have recognized some or all of these phenomena. Perhaps you have a deeper feeling within yourself of apprehension, of concern, of uncertainty, a growing uncertainty over whether things are really going to be all right. There is the appearance of normalcy, of course, but the feeling of change is pervasive.

If you stop to ask yourself: "How do I really feel about the condition of the world?" or "How do I really feel about the condition of my country and the welfare of the people?" you will evoke perhaps a different response than merely trying to assume that everything will be fine and that the change at hand is merely a fluctuation or a minor disruption and that life will return to normal as soon as these problems pass over, as if they were clouds just passing over the land, where you just need to weather the storm, a brief but difficult storm.

Yet what gives rise to a deeper sense of apprehension is not a brief disruption, is not an inconvenience. It is something far deeper and greater, whose impacts are far reaching. This is what gives rise to a deeper concern and sense of apprehension.

It is important here not to dismiss these feelings, for they are signs within yourself that you are responding to something greater in the world even if you have not recognized fully what it is that is generating this response within you.

For at this moment, the world is giving you signs, and Knowledge within yourself—the deeper Intelligence that God has created and put there for you—is also giving you signs. The signs are not merely forms of encouragement, or affirmation, or confirmation of your ideas, beliefs and expectations. More often than not, they are warning signs—alerting you to the presence of danger, or indicating some change in your understanding or activities.

It is as if you had an early warning system within yourself, giving you signs and clues that you must pay attention to something, that you must respond to something, that you must recognize something that previously you had not recognized. This early warning system is vitally important for you and should never be construed as merely being fearful or negativity on your part. It is part of the greater Intelligence that God has put within you to warn you of the presence of danger, or to alert you to a need, to give your attention to a specific thing or a set of difficulties.

Look at nature. The animals in the field are always looking about themselves to check their environment, to see if there is any risk or

ESCAPING FEAR, CONFUSION AND HOPELESSNESS

danger there. Why is it that they would show such caution and not you yourself? Wisdom would instruct you to always be watchful and alert and present to your surroundings.

Every moment of every day people are making critical mistakes, or having terrible accidents, or overlooking important opportunities because they are not paying attention to their environment. They are not paying attention to their deeper inclinations, which are alerting them, encouraging them, or in some cases holding them back. This lack of attention then is a critical problem and is the source of the vast majority of mistakes and errors, and even catastrophic errors, that people make.

Do not think that to be so careful and to be so observant is to be acting in a state of fear. It is really functioning in a state of wisdom. For the wise man or woman is always careful to discern their environment, to listen more deeply for the motives and intentions of other people, to check themselves in making conclusions and in approaching decisions very carefully, seeking verification from others whom they trust.

It is this care and caution that represent wisdom. It is not fearfulness. This wisdom is born out of the recognition that the world has many hazards and many opportunities, and one must be observant and present to see them both. To be reckless and inattentive is to put oneself in jeopardy and also to miss the critical signs, the critical opportunities, that life is giving you so very frequently.

Therefore, to be cautious and observant is really to be intelligent, and to show a lack of awareness and a lack of caution demonstrates a lack of intelligence. For what is intelligence but the willingness and ability to learn and to adapt? Real intelligence creates the desire to learn, to be fully engaged with life, to be watchful, to be discerning, to refrain from premature judgment or conclusions, to give your attention fully to things in front of you, to listen carefully to other people to discern their motives and intentions and the real communication that they are attempting to give through their words and through their actions. Intelligence is not simply being clever or solving complex problems. Intelligence is the willingness and ability to learn and to adapt.

Life is a constantly changing situation, presenting both hazards and opportunities. To not be fully present to that is to not be present to oneself, is to be functioning in a way where you are disengaged or distracted from this primary engagement with life. In that situation, you cannot bring your greater powers of mind to bear. You cannot use the greater Intelligence of Knowledge that God has provided you, and you will miss the signs of the world and even the signs from Knowledge itself—guiding you, telling you, instructing you, encouraging you, setting you forward or holding you back, depending upon the situation at hand.

It is this lack of attention, then, that disables you from gaining the full benefit of being in the world and from utilizing this greater Intelligence of Knowledge, and in fact from utilizing all of your skills, all of your levels of sensitivity. It denies your creativity and your ability to bring new ideas or understanding, new creations into the world.

For if you cannot respond to the world, then you will not respond even to your creativity. It is as if a fog has come over your mind, or you have drawn a curtain between yourself and everyone and everything else. And the only way you cannot be present to the world is to be obsessed with your thoughts and fantasies, or the thoughts and fantasies of others.

ESCAPING FEAR, CONFUSION AND HOPELESSNESS

Your intellect is a wonderful tool, a tool of communication in the world, and that is its ultimate purpose. But the mind that does not have this purpose, that is not guided by Spirit, or Knowledge, becomes within itself a form of obsession—pulling you in, engaging you endlessly in self-debate, self-doubt, the projection of one's desires, or the nightmare of one's fears.

It is a great tool here that has not been used properly, that is not being applied correctly, that has no governance because ultimately it is your body that must serve your mind, and your mind that must serve your Spirit. Without this service, the vehicle of the body and the vehicle of the mind become reckless, become problematic and become dominating, dominating problems.

When you are out in the world, you must really pay attention. Do not be listening to your music when you are walking about. Watch. Listen. Deepen and cultivate your seeing, your hearing, your touching. Do not be consumed in your thoughts, your problems, your memories or your concern for the future. This is being present. This is being alive. This is really being intelligent.

You do not need a great education in the usual sense to be intelligent, but you do need to be observant. And you do need to learn to gain wisdom from your experience and from the experience of others. It is utilizing the presence of Knowledge, which already exists within you, and it is gaining worldly wisdom that represents the evolution of intelligence—both within yourself as an individual and for all of humanity.

Whereas people have particular views and come to particular conclusions, Knowledge and wisdom both have a universal resonance about them. People may disagree on how to fix something, or how to interpret something. They may disagree about their ideas about what is correct human behavior or how governments should be formed, or how commerce should be undertaken. There is a great variety of opinion here, a great range of interpretation.

But real wisdom is something that people can share readily with one another, and the experience of Knowledge is something that unites people. For there is no your Knowledge and my Knowledge. There may be your interpretation of Knowledge and my interpretation, or the range of your experience and the range of my experience. But Knowledge itself is one. That is why it is a great peacemaker in the world. That is why it is a medium through which people recognize each other, resonate with each other and find meaningful action in relationship with each other.

What is fear? Fear is a rejection of reality. It is a projection of one's own inability to function successfully in life. It is the result of living without the guidance of Knowledge and without accumulating sufficient wisdom.

There is fear in the moment in the face of real danger. That is normal. But most of the fear that consumes people is a product of their imagination. They are dying a thousand times in their imagination. They are imagining all kinds of terrible possibilities, things that could happen to them or to others, to their loved ones, or to the whole world. They are living their own nightmare, but it is a nightmare of imagination.

It is the result of being disconnected from life and from the deeper Knowledge that God has placed within you. Without this greater certainty and without this greater engagement with the world, now you are alone. Now you feel vulnerable. Now you lack certainty. Now you are feeling stressed because you do not know what to do and you have not built a foundation within yourself upon which you can rely,

ESCAPING FEAR, CONFUSION AND HOPELESSNESS

a foundation of self-reliance, a trust in Knowledge within yourself and with it the ability to be fully engaged in the world.

People who are fooling around a lot, not being attentive, not paying attention, carried away in their own streams of imagination and self-concern, are ill prepared to deal with the realities of the world and are certainly ill prepared to deal with the Great Waves of change or with the reality that humanity is now in contact with intelligent races from the universe.

This inability, then, to be present, to learn from life, to recognize life's hazards and to learn from its teachings and opportunities, this represents a great handicap. And within this handicap there is endless fear and apprehension, but not apprehension that comes from Knowledge, but more apprehension that comes from one's own weakness and lack of discernment, lack of experience and lack of self-confidence.

If you have not yet built a sufficient foundation of strength within yourself—strength representing your connection to Knowledge and the building of worldly wisdom within yourself, strength built from having meaningful engagements with other people—then you will look at life with trepidation, and you will feel trepidation because you are not yet able to deal with reality. Having given yourself so much to fantasy, having allowed your mind and imagination to be captured by the imaginations of others, your real ability to be in the world and to navigate the ever-changing conditions of the world is undeveloped. And without skill, you will feel helpless.

The only way to really avoid fear and all of its dark manifestations is to build your connection to Knowledge and begin to assemble your worldly wisdom, and distinguish them from your beliefs, your wishes, your fears, your political ideology, your religious ideology, your attitudes, your prejudices and so forth. This involves real work and effort on one's own behalf, real inner work to gain a greater control over your mind and to utilize it for a greater purpose rather than simply being a slave to all of its fantasies and all of its nightmares. It represents one taking power within one's life instead of floating along in a sea of assumptions, drifting in one's own imagination.

Imagination itself is very valuable if it is guided by Knowledge within you. Your intellect is tremendously valuable if it is guided by Knowledge within you. But without this guidance, these things become errant. They are out of control, and they generate hazards and self-obsession.

The world is teaching you what is coming over the horizon. It is showing you through demonstration all of the errors that people are making. It is showing you through demonstration all of the ways that people are successfully engaging with the world. It is showing you by demonstration what real contribution to the world looks like and how it can be experienced. The world is showing you every form of human error and misinterpretation.

But to gain this great education from the world, you must look with clear eyes. You must look without drawing early conclusions. You must live with questions instead of always seeking for answers. You must be present to life instead of just present to your own ideas.

Taking a beginner's approach here is very helpful because a beginner makes few assumptions about his or her own learning or accomplishments and is present to learn through experience and through demonstration what is most valuable, and to distinguish what is valuable from what is not valuable.

ESCAPING FEAR, CONFUSION AND HOPELESSNESS

Because fear is the result of not being present to the world and not experiencing one's connection to Knowledge, the deeper Mind, then the antidote to fear is to become present to the world and to take the Steps to Knowledge within oneself.

The closer you are to Knowledge, the more you can gain objectivity and the less you are swayed by the fantasies of hope or the nightmares of despair. You have found a way to be with the world that is clear and discerning and compassionate. You are able to look and see because you are not so afraid. You are able to face yourself and your own problems because you are not so afraid. You are able to look out on the horizon of life to see what is coming over the horizon because you are not so afraid. You are able to find the truth within yourself and to distinguish it from everything that is untrue because you are not so afraid.

It is this willingness and ability to adapt and to learn that gains such strength within you, that gives you this kind of objectivity and this freedom from fearful imagination and from the seductions of fantasy. It is a clarity, but it is a clarity that rewards you. It frees you from self-doubt. It frees you from self-repudiation. It enables you to look at the world with clear eyes to discern its meaning, its value and its difficulties.

In this way, you are able to navigate not only the changing circumstances of your life, but greater waves of change that are coming to the world, which will produce ever-greater degrees of instability and uncertainty in all nations of the world, even the wealthy nations.

How are you going to know what to do in the face of this? How are you going to know what to do when circumstances change, when you have to make difficult decisions in the moment, when you have to

PREPARING FOR THE GREAT WAVES OF CHANGE

rely upon Knowledge within yourself instead of simply following the edicts or the predictions of people in positions of authority?

You must assume a greater responsibility for your life. That is fundamental. This, then, gives the encouragement that one needs and the commitment one needs to bring oneself fully into the moment and to discern one's assets and one's liabilities, the strength of one's relationships, the strength of one's connection to Knowledge, the appropriateness of one's own circumstances.

Are you in the right place with the right people doing the right thing? Are your relationships encouraging you or discouraging you? Are they bringing you closer to reality or are they taking you away? For there are no neutral relationships. They are either helping you or hindering you in finding and building your strength, cultivating your wisdom.

In a world facing the Great Waves of change, these things now become ever more important. If you are idly sitting on the sidelines of life, preoccupied with television or movies, living out someone else's imagination, disconnected from the events of the day, unaware of the great changes that are building in the world, then you are putting yourself in a position of extreme powerlessness and vulnerability.

As a result of this lack of awareness and preparation, you will feel immensely afraid. And as the Great Waves build, your fear will escalate. You will feel hopeless and helpless and be prone to panic and to making very bad decisions or to following others who are making very bad decisions.

God has put a greater Intelligence within you to guide you and to protect you, but you must gain access to it. You must listen to it and

ESCAPING FEAR, CONFUSION AND HOPELESSNESS

learn from it and refrain from making self-comforting assumptions about life and about your future.

Do not think that everything will work out fine. Everything will work out, but perhaps it will not be fine. Do not think that God will take care of things in the world, for God has sent you to take care of things in the world. Do not think that everything happens for a good purpose, for things just happen, and yet they can be used for a good purpose. Do not think you are at the right time and the right place with the right people, for this is a false assumption. You may be in the wrong place with the wrong people doing the wrong thing, and that is actually very common for people around you. Do not think that all fear is bad, for there is fear that is born of caution, the recognition of real danger, requiring a serious and wise response.

Learn about the Great Waves of change, and look out into the world with as much objectivity and courage that you can muster. Gain over time the strength to do this—looking every day, contemplating every day, seeing what change you must bring to your life and circumstances, avoiding complacency and ambivalence—these two plaguing maladies of the human mind. Avoid assuming either self-defeat or self-success, and venture into life without these false assurances or predictions. Face difficulty with the power of Knowledge within yourself. And serve others, not because the outcome is guaranteed or assured, but because the service is needed, and it is correct for you to provide it.

Fear is easy. It is easy to become fearful, to fall into the pit of fear or despair. It is easy to feel helpless and hopeless. It is a lazy response to life and to one's circumstances. It is more difficult to face the challenges and to meet the difficulties on the road of life, to examine everything that must be examined, to fix or repair the things that must be fixed or repaired, to face your weakness and know that

PREPARING FOR THE GREAT WAVES OF CHANGE

you must become stronger. Do not give in to the dark path of acquiescence or self-assumption, or hopelessness and despair.

In the face of the Great Waves of change, and indeed in the face of Intervention from races from beyond the world, people will become immensely afraid and will fall into despair. They will do this automatically because they have no foundation to face life with strength and objectivity.

Instead of facing the situation and asking, "What must I do?" and then considering what they must do, they will simply fall apart. Or their hope will evaporate as if it were a phantom. They will fall from self-assurance into the pit of despair as if they were falling out of a building. They do not have the strength or the foundation within themselves or the connection to Knowledge to face an unanticipated situation, an unexpected event. They will go into denial, and they will turn away, not because they are facing something that is unreal, but because they do not have the strength to face it. They cannot handle it. They cannot handle it because they do not have the strength, and they do not have the foundation, and they do not have the connection to Knowledge.

In the absence of these things, there is only hope and fear. And hope is easily defeated. And fear is easily adopted. It is as if you have nothing to stand on in life except assumptions and beliefs, and when those assumptions are threatened or proven to be false, you fall right out of the sky. Whatever was holding you up before has deflated, and now you are falling out of the sky—falling into despair, confusion and panic.

In facing the Great Waves of change, you will see this in people, this inability to face the situation, this inability to be with the problem, this demand for simple solutions, this demand for someone else

ESCAPING FEAR, CONFUSION AND HOPELESSNESS

to take care of the problem for them. You will see this. You can see it even now.

Fundamental resources in the world are diminishing. That is why the cost of everything is going up, but who can face that and see what the real cause is? Even the experts cannot do this. So governed they are by hope and ambition, they cannot see the reality. Life is telling them what it means, but they cannot see it or hear it. It is as if they are blind.

Formal education does not seem to make much of a difference here. The educated do not seem to have much more vision than the uneducated. They just interpret things in more complex ways.

You either see the signs of the world, or you do not. But to see them, you must look and keep looking. If you do not look, you will not see. If you do not see, you will not know. If you do not know, you will have no sense of where you are in life and what is happening. And you will not have the strength to deal with the changing circumstances that are bound to occur.

You may speculate about life and change. You may bring a spiritual perspective and think that everything is being governed by a great plan, but it means nothing if it only hides the fact that you do not have the strength to face the reality of your life and to recognize the Great Waves of change that are coming to the world and the strength to prepare.

Your spiritual ideas mean nothing if you have not built a foundation in Knowledge and the courage to face things and problems that cannot be easily answered. Some problems do not even have answers. Great problems do not have simple solutions. You must live with the problem and work with the problem to find ways of resolving it or removing its hazards. People who want simple solutions want this because they are not strong enough to face the situation.

Fear here is a product of a lack of preparation on one's part and the result of one not using their intelligence, not gaining worldly wisdom. This is different from being concerned. Concern is recognizing risks and difficulties. Fear is being helpless and hopeless and overwhelmed. It is the difference between one getting off a sinking ship and one simply being frozen, not knowing what to do.

It is important to look at fear in this manner—the fear born of a lack of strength and preparation, born of the lack of having a real foundation of wisdom and strength, fear born of the absence of Knowledge in one's awareness, fear that is the product, the inevitable product, of one living upon a set of unquestioned assumptions and beliefs.

God has given you a greater Intelligence to navigate the difficult and uncertain times ahead, but if you have not gained access to this Intelligence and brought it to bear in your life, then how can it possibly be of service to you? Without this deeper certainty, you only have your assumptions and your beliefs to give you any sense of stability and assurance. And what are assumptions and beliefs but just ideas in the mind? They may have nothing to do with reality at all.

The vast majority of mistakes and tragedies that people face are the product of their not paying attention, the product of their not being present to the situation, the product of not checking something out with Knowledge before making a big decision, of not heeding the warning signs, of not seeing the power and the grace of Knowledge working on their behalf.

ESCAPING FEAR, CONFUSION AND HOPELESSNESS

Fear occupies the emptiness, the emptiness that is the result of one not building one's foundation sufficiently. Instead of a foundation, there is nothing there, and within that space, fear, anxiety, apprehension, self-doubt begin to fill the environment. Where reality does not exist, imagination takes over—imagination that does not have any real guidance or focus to it.

Here people fantasize and dream and speculate, but really they do not know where they are in life and they do not know what they are doing. They are unprepared for what is coming over the horizon that could be discerned if they were looking, if they were paying attention and if they maintained their attention.

You were sent into the world to serve a world in transition, a world that will be facing the Great Waves of change, a world that will be facing the tremendous threshold of engaging with intelligent life beyond the world. You will be facing a world of declining resources, of economic instability, of political and social discord. You were sent into the world to face these things and to provide service to a struggling humanity. And yet is this your experience of who you are and why you are here? And do you really have the strength and the ability to do these things?

If you do not, after an honest reckoning, if you realize you do not have this strength or this awareness, then you can see what is necessary for you to focus upon: to begin to build this strength, to gain an education, to learn about the Great Waves of change, to take the Steps to Knowledge, to sober your life so that you can be present for it, so that it can be of great benefit to you, to set aside all these forms of personal avoidance and escape, to turn off the television set, to stop reading foolish things that have no meaning and to begin to focus on developing a wisdom and an awareness about what is happening in your world, to take the time to discern and to clarify and to write out your strengths and your weaknesses so you know what they are.

Where are your blind spots? Where do you tend to be foolish? Where do you tend to believe in other people without really questioning the value or the meaning of what they are telling you? What is the state of your personal health? What is the state of your economic situation? Are you in a position to deal with great change, or are you extremely vulnerable? Where are your strengths and your weaknesses, your assets and your liabilities? All these things then become a focus now, and they are all an antidote to fear and to negative imagination.

If you have not built your foundation, now is the time to start. Doing something really important for yourself will give you a growing sense of self-confidence. Taking action, not simply being aware of something but taking action regarding it, will give you a sense of confidence and movement in your life. You need this confidence, and you need this movement. You need to see that you can actually do something. Even if it seems insignificant, you can do something, and you are doing something, and you are keeping your eyes fixed on what it is that you must prepare for. You are avoiding all the tendencies within yourself to want to give up or to run away or to go into denial and avoidance. This is filling your life with strength and certainty, creating very little space for fear to emerge.

Do not worry what other people are doing. It is what you must do. Learn from those who are doing more than you are doing, but do not wait for a consensus, for the consensus will not be a consensus of wisdom. Do not follow the weak. Do not identify with the weak. And do not condemn the weak.

Instead, build your strength. Build your foundation in Knowledge. Bring your life into order and do not stop. And instead of fear, you

ESCAPING FEAR, CONFUSION AND HOPELESSNESS

will begin to feel the Power of God working in you and through you. And regardless of how difficult the circumstances may seem around you, you will be able to bring this strength with you.

This is your gift to the world, the world you have come to serve. It requires you to become strong and capable, aware and observant. It requires that you escape fantasy and denial, helplessness and hopelessness, for you are here to serve the world. And those who sent you into the world are your strength and your reminder that you are here for a greater purpose.

As revealed to Marshall Vian Summers, June 25, 2008

CHAPTER 7

Who Can You Trust?

In times of great change, many things are revealed. Weakness is revealed, corruption is revealed, incompetence is revealed—within governments and institutions and within people everywhere. In facing great change, people begin to find the real condition of their nature and constitution, and whether their foundation is strong or not.

Humanity is now facing Great Waves of change that are coming to the world: environmental decline, diminishing resources, violent weather, changing climate, growing political and economic instability and the real risk of war between groups of nations over the remaining resources.

The growing human family is drinking from a slowly shrinking well. The strain this will produce—the unanticipated consequences, the fracturing of economies—all of these things will be revealed as time goes on.

People will find themselves far less secure than they perhaps believed they were before. It will change their priorities and their focus in life, and the question of trust will arise. Who can you trust? What can you trust? Can you trust yourself? Can you trust other people? Can you trust your ideas or beliefs or assumptions?

PREPARING FOR THE GREAT WAVES OF CHANGE

So much will be thrown in question that you may feel you cannot trust anything, that everything is in flux. What you relied upon before is now changing or breaking up or falling apart. You may feel that you cannot trust your leaders, your government, your economic institutions, your religious institutions. People may question the foundation of their religious beliefs and institutions. Peoples' grievances towards other groups and other nations will be activated as they look for someone to blame.

You are entering into a time of great instability and uncertainty, and this will arouse many different kinds of passions, ideas and reactions in people. Most of them will be unhealthy and unfavorable. Within yourself, you will have to find a firm foundation as the foundations of society and organizations around you become challenged or fall into disarray.

Trust here cannot be based upon hope. Hope is too weak and too easily dashed, too easily disappointed and overthrown. Your trust must be in something greater, within yourself and within other people.

As the Great Waves approach and begin to have ever-greater impacts, you will see, perhaps with dismay, that so many things you trusted in before or relied upon before will now appear to be weak and vulnerable. People will appear to be weak and vulnerable. Governments will appear to be weak and vulnerable. Even your most firmly held beliefs about your nation or your religion may appear to be weak and vulnerable. This throws you back upon yourselves and requires a re-evaluation, a very serious and deep re-evaluation.

Many people have based their entire lives on a set of hopes and expectations which are built upon a weak and fragile infrastructure of human society. You will see how easily governments are thrown

WHO CAN YOU TRUST?

into panic, how easily major institutions can fail or falter. And this will shake you.

Perhaps you will want someone to blame. Perhaps you will strike out at others. Perhaps you will express anger and frustration, but this is not really addressing your core need or problem here.

The question arises: Who can you trust, what can you trust? Everyone will trust something. Even if it is a distrust, they will trust something. Perhaps they will trust change. They will say everything is going to change, and that is where they will place their faith and their trust. Even the most cynical point of view has trust in it. One is trusting their cynicism. One is trusting their ideas. One is trusting their perception and their position in life, even if that trust is misplaced.

So the question will arise in times of great change, uncertainty and upheaval: What can you trust? Where will you place your faith? As you look about you, if you are looking seriously, if you are really looking to see and not merely to judge, it will be a difficult question to answer. What can you really trust? What is it you see that will not fall away in front of you, or prove to be weak and vulnerable or prove to be other than it appears?

It will be a crisis of faith for people everywhere, even in the wealthy nations where hope and belief and assumption are more rife and evident than in the poorer nations, where people tend to be a little more realistic. Fantasy and self-indulgence increase with affluence, but so does disassociation from life—a disassociation from humanity, a disassociation from reality itself.

God has created a deeper Mind within you—a Mind that is not afraid, a Mind that is not in conflict with itself, a Mind that does not judge and condemn others, a Mind that is secure, a Mind that can face great change and upheaval with determination and equanimity, a Mind called Knowledge. It is called Knowledge because it is related to your ability to have profound experiences of recognition and knowing. These experiences can fly in the face of even your own ideas and beliefs, for they are not bound by these things.

Your deeper Mind is so different from your intellect; from your socially-conditioned mind; from your personal mind that is patterned by culture, family and even religion; patterned to think like others, patterned to be compliant, patterned to be subject to other forces; to be persuaded and manipulated, to be corralled and cajoled, to be directed and to be pacified.

That is why people say the same things, think the same thoughts, do the same things. Like herds of cattle, they are corralled around, led to believe this, led to believe that, assured constantly by governments and leaders that their well-being is being attended to, that they will be safe and secure.

But in the face of the Great Waves of change, these assurances will prove to be weak and transparent. And they will prove to be utterly false in so many instances. Then people will be outraged. They will be shocked. They will be dismayed. They will feel betrayed. They will feel misled. And they have been misled, and they have misled themselves.

Having given their authority and their power over to other forces, to other people, to governments, or to sets of beliefs and assumptions, now they feel really vulnerable. And they are less likely to trust future assurances.

This is both dangerous and healthy. It is dangerous because people can do radical things that are not in their best interests. They can condemn others, trying to find someone to blame because they

WHO CAN YOU TRUST?

themselves cannot yet take real responsibility for their life and condition, so they must find someone else to blame. And they will give this blame with vehemence. Perhaps they will blame peoples of other nations, leading to further misunderstanding and the potential for conflict and war. This is how your personal mind responds to change—with fear, with panic, with aggravation, with condemnation, even with violence: violent thoughts, violent behavior, violent intentions.

But your deeper Mind does not respond this way at all. It recognizes the Great Waves of change. It understands people's lack of preparation and awareness. It sees the problem. It recognizes the manifestations of this fundamental problem that people are not connected to Knowledge within themselves, so they do not see, they do not know and as a result they do not act wisely.

They allow themselves to be shepherded around by people who are profiting off of them. They are given some slight degree of security and affluence perhaps, but they yield so much for this. They are willing to believe because they want to believe, because they are afraid not to believe, because they feel they do not have the power to think outside of these assurances. This is fundamentally a problem in people's lack of awareness of Knowledge.

Would you rather have a government guide you, or a financial institution guide you, or have God guide you? Your leaders may be as blind as anyone else, living on false assumptions, believing firmly and fervently in their own ideas and the justification of their actions, like the blind leading the blind. You want to believe because you are afraid not to believe, because without that, then what can you trust?

This shock is beneficial because it gives you a chance to recognize how weak and frail is the foundation of human civilization and how prone to error all people are. And you will see that there is really something else within people—perhaps you will only see it in a few people—that is really wise.

You have a natural trust towards people who demonstrate this wisdom and this compassion. These people are not perfect, of course. They can make critical errors, but there is a certain quality about them that distinguishes them from others and that naturally evokes your sense of respect and trust. Though these individuals may make mistakes, they are functioning from a higher standard. They are truly here to serve, and this transcends their self-interest in the matter. This is because you are recognizing Knowledge within them—a deeper power, a greater incentive, a truer motivation.

To be disillusioned and disappointed in others and in leaders of governments, commerce and religion is healthy because it requires you to reconsider your own beliefs and assumptions and to consider that there may be a deeper power within you that is far more reliable and consistent. This opens the door. It is a possibility that you will have this recognition. It is not assured.

In times of great change and upheaval, people will do very foolish things. They will hurt other people. They may even hurt themselves. But there is a chance that they will see that there is a deeper power and presence within them that is not a product of social conditioning or social patterning. It is not a product of culture, belief or tradition.

Here there is something very unconventional within you—a free Mind, a Mind that is uncorrupted, a Mind that is beyond persuasion and manipulation. This is not the mind you think with every day. This is a deeper Mind.

WHO CAN YOU TRUST?

People say they trust themselves. Perhaps they will even declare that they will only trust themselves. But what are they really trusting within themselves—their ideas, their beliefs, their assumptions, their political views, their religious beliefs?

What are they referring to here? Are they going to trust their emotions, which can be so easily stoked by fear, disappointment and blame? Are they going to trust their natural instincts, which in many cases can be truly violent and destructive? What are they going to trust within themselves? What are they talking about here when people say they are going to trust themselves?

This can be just as foolish and illusory as putting all your faith in some kind of institution or set of government leaders whom you really do not know, because it is putting the faith in the same shallow intelligence. To believe in your own social conditioning is really no different than believing in someone else's social conditioning. It is belief in the same thing.

Perhaps you feel you have more control over your social conditioning, but most people really do not. They are simply following it slavishly, and when it disappoints them, they try to place the blame outside themselves. To avoid self-recrimination, they try to blame others to find someone beyond themselves upon whom they can vent their anger and frustration.

To say that you trust yourself would only really be wise if you were speaking about Knowledge, but Knowledge exists beyond the realm and the reach of the intellect. It cannot be understood intellectually because it is a force and a power that is more ancient and more permanent than your personal mind. It is like asking a child to understand the reality of their parents.

PREPARING FOR THE GREAT WAVES OF CHANGE

Knowledge is functioning at a different level. It is not of this world, but it is in this world to fulfill a mission, to experience and express a higher purpose. This is your mission and your higher purpose. In light of this, disappointment and radical change, no matter how difficult or stressful, can really be beneficial. If seen from the position of Knowledge, then the Great Waves of change are the perfect opportunity to express a greater purpose and to give greater gifts to humanity.

Yet you look with sobriety because you understand the great risks for human suffering and loss of life here, so you do not take this casually. It is very risky. Change, if it is real, is very risky. The outcome is not assured. People may not do well or even survive. Knowledge is very sober about this. It is not casual. It is not so self-assured that it can assure the outcome, but that is not its issue or its emphasis. It is here to serve.

Because it is immortal, it is not afraid of death. It simply does not want to lose this opportunity to serve. After so much has been done to bring you into the world, and after so much influence has been placed in your life to keep you from making terrible mistakes, or giving your life away prematurely, or destroying your opportunity to contribute to a world in need, Knowledge definitely wants to preserve you—to preserve your mind, your thinking mind, and your body. In this regard, it will protect you and guide you with the wisdom you will not be able to find anywhere else.

Knowledge looks upon the world very differently. It sees the need for Knowledge. It recognizes that even in affluence, people can be betraying themselves constantly and undermining their future. It sees beyond the illusion of success and failure. It even sees beyond the illusion of love and fear because most people do not understand what love and fear really are. It sees Knowledge and the need for

WHO CAN YOU TRUST?

Knowledge in the world. It looks with compassion. It looks with wisdom. And it looks without self-deception. It sees the problem. It sees the risks. It sees the needs. It is always watching. It is not caught up in fantasy. It is not distracted. It is not carried away by promises of love and bliss and enlightenment. But it is also not terrified by the world in its brutality, suffering and degradation.

Your personal mind cannot think like this. You may try to purify yourself. You may give yourself over to decades of spiritual practice, but your personal mind cannot do this. It is still a product of the world.

People try to lift their intellect beyond error, beyond illusion, beyond misinterpretation, but they find they can only go so far. And then they have an experience of judgment or condemnation, fear or blame, or lust or desire, and they realize that really it is the same old mind. Perhaps they have become more objective. Perhaps they are less influenced by this personal mind, but it is the same old mind.

Just like you cannot alter your body really without extreme actions, without mutilating yourself, you cannot really change your appearance beyond a certain point. You cannot change your personal mind beyond a certain point. What changes is your relationship with it.

Instead of slavishly following your thoughts and your feelings, thinking that your mind is who you are, you begin to look at it more objectively. You see that your thinking mind—your thoughts, your beliefs, your attitudes—really is not who you are. It is an outward expression of who you are, but who you really are is something more deep, more profound.

PREPARING FOR THE GREAT WAVES OF CHANGE

It is like looking at the surface of the ocean—one day calm, next day turbulent, always changing, always being swept by the winds of the world. Do you look at the surface of the ocean and know the ocean? Of course not. You cannot see the life the ocean contains from the surface. Perhaps you will see the evidence of a few life forms—whales and dolphins and fishes—but you cannot see what really lives in this ocean. You cannot stand on a mountain top and look at the forest below and know everything that lives there, unless you have lived there. Do not look at the mind, at the surface, and think you understand the mind.

There is Knowledge deep within you. It is the deeper Mind within you, a Mind so different from your intellect that you cannot even compare them. This Mind knows. This Mind sees. This Mind waits. This Mind directs.

For you see, God knows the world is extremely difficult, and living in Separation is extremely hazardous. That is why God has placed Knowledge within you to guide you, to protect you and to lead you to discover your greater purpose in the world—within the very circumstances of the world as it is and as it will be.

You are not sent into the world to set yourself up in a state of perpetual comfort and self-assurance. You are sent into the world to work, to accomplish things, to make deep connections with others. To seek things beyond this or to think things that contravene this is a kind of self-betrayal.

And yet does not your culture encourage this? Does not your society value this, this self-betrayal—encouraging you to be powerful and wealthy, to have pleasure, to live in great affluence, to be admired, to be feared?

WHO CAN YOU TRUST?

Society's visionaries and saints may be reduced to poverty and persecution, but oh its wealthy are idealized and held in great esteem. Those who have lost their connection to Knowledge and have given themselves over to seeking power and preference in the world, is that who you would emulate? Is that what you wish for yourself?

Certainly, you need a certain degree of security. Basic needs must be met. You need a stable environment. You need opportunities. Yes, of course. Knowledge knows that and supports that. But really people have taken this way beyond their authentic needs.

So you have an aberrant life here, a life that is perpetually unstable, a life that is driven by fear and greed. It is a distortion, and in some cases it is an abomination. In times of great change, you realize finally you cannot trust this. Perhaps your belief in it will be shaken. Perhaps your slavish adherence to it will be shaken and thrown into question and doubt. And this is a good thing, you see.

In times of seeming affluence, people are asleep. Few are questioning their condition or the validity of their lives or the meaning or value of what they are doing. But in times of disappointment, people are much more alert, much more concerned. This generates for many people a healthy self-inquiry and re-evaluation.

If this re-evaluation is successful, it will bring you to Knowledge. If disappointment is beneficial, it will bring you to Knowledge. If self-doubt is productive, it will bring you to Knowledge. If disappointment will yield real value to you, it will bring you to Knowledge.

You do not understand Knowledge. Perhaps you think it is weak and fleeting, a kind of romantic idea, a kind of hope or a wish. Perhaps you think it is not substantial enough to really be your foundation in life. You still value your technology. You still value your institutions and beliefs and ideas, thinking they will save you and the world.

But look at the individuals who had the greatest beneficial impact on humanity over time, and you will see that they were individuals who were guided by a deeper and more pervasive power and a truer intention. And they demonstrated a greater courage and integrity. They were guided by Knowledge, just as you must be guided by Knowledge.

How can you trust Knowledge? You barely experience it. It is not there when you want it. It does not give you what you want. It does not answer all your questions. It does not reassure you the way you want to be reassured. How can you trust this thing called Knowledge?

Like anything else, real trust must be based upon experience. It cannot simply be a hope or a wish or an idea. You build your trust in Knowledge by following Knowledge and seeing how it brings into your life people and situations, opportunities and insights that you would not have found otherwise.

You learn to trust Knowledge because you realize that it is not fooled by the world and that somehow you had feelings or premonitions about things before they happened. You felt restraint before you made mistakes. You felt encouragement, but then did not follow it through. You chose your preferences, which are all based on fear, the fear of not having. You followed this when your deeper nature was telling you something else. You wanted to own this thing, but Knowledge within you was silent, showing no interest at all.

Over time and experience in taking the Steps to Knowledge, you begin to see the very great difference between how Knowledge responds to things and how your intellect and mind respond to

WHO CAN YOU TRUST?

things, and how much energy you spend over things of little or no value or consequence, pursuing things of little or no value or consequence—your emotional rages, your outbursts of anger, your indulgences in self-pity. All these things are happening, and Knowledge is just there.

You see the contrast here that your personal mind is like the surface of the ocean—always turbulent, always changing, always being thrashed about by the world and by external forces. And then there are the deeper waters, which are really moving the oceans all over the world. And there is greater life beneath the surface.

This contrast over time will show you beyond your doubt that there is a greater Intelligence within you that is really not functioning at the same level as your personal mind, and that you are burning up your life force in reaction to things on the outside when in reality there is a deeper awareness within you that is not really affected by these things. It is watching and waiting. It is trying to keep your life on track, to keep you from going too far off course, to restrain you from giving your life away to people, places or things before you even know what your life is.

How many people on their wedding day have entered the ceremony with a great feeling of inner restraint—Knowledge trying to keep them from giving their life away before they even know who they are and where they are going? How many people have capitulated to the intentions of others in this regard? The pressures of their family; the expectations of their parents; the proclamations of their religious leaders; the overall pressure of their cultures; pushing them into a life that is not really their life; pushing them to believe, to accept and to acquiesce to a set of circumstances that does not represent their real destiny and purpose in the world.

PREPARING FOR THE GREAT WAVES OF CHANGE

The tragedy of this is everywhere, amongst the rich, amongst the poor, everywhere. You know what this is, for you have been subject to it as well. It is the real dilemma of your life, you see. It is the battle within yourself.

Knowledge is not fighting. It is simply going in the direction it must go. It has a destiny here in this world. It is your destiny—a destiny to meet certain people, to accomplish certain tasks.

Do not think that what you are doing now and the people you are with now represent this destiny, for in most cases, that will not be true. Do not use this as a kind of justification for your current involvements.

You may convince your mind, your thinking, your feelings because you want confirmation, because you do not want to face the fact that you have made many mistakes and errors in judgment. But you cannot persuade Knowledge. And you cannot persuade anyone else who is strong with Knowledge. They will see right through it. Just like they see right through every clever and ingenious device that people create for themselves or accept from others to try to gain advantage.

This is the power of Knowledge. It cannot be fooled. And, after a while, particularly in the face of the great change, you yourself realize you do not want to be fooled. You do not want to give your life to things that have no substance, no stability, no future. You do not want to face that kind of disappointment and frustration anymore.

Here disappointment can really serve you, to bring you to Knowledge. Here being honest about your feelings, rather than pretending to be happier than you really are, will really serve you. It is the beginning stage of liberation. To keep believing in the same

WHO CAN YOU TRUST?

things that are not real is just to recommit yourself and your life to further disappointment.

Knowledge is calling to you. It is advising you every day. It is not talking to you like your mind talks to you, like you talk to yourself. It is more like a force of attraction. It is given to certain things and not to other things. It sets a direction. It is an influence. It is a power.

Your mind, your thinking mind, can be talked in and out of a million things. But Knowledge just has direction. Perhaps it will say to you: "Do not do this thing. Do not say this thing. Do not commit yourself here. Follow this. Give yourself to this." That is the extent of its conversation. It is not a little chatter box constantly talking away, trying to reassure itself, trying to be accepted, trying to impress others. It is so very different you see, refreshingly different.

You were sent into the world for a purpose. Knowledge holds that purpose. Knowledge drives that purpose. Knowledge is here to keep you from giving yourself away to other things, to committing yourself prematurely, to making critical errors. It is here to guide you, but you must come to Knowledge. You must learn to abide with Knowledge, build a connection to Knowledge, take the Steps to Knowledge.

It is your thinking mind that must yield here. Knowledge is not here to be your little servant. You are here to serve Knowledge. That is the difference, you see, that will make all the difference in your experience and in what you will be able to see, know and do.

At the outset, people treat Knowledge like a resource. "Well, I just want to have insights when I want them, so I'll just go and ask Knowledge," as if Knowledge is there as some kind of a personal counselor—a resource.

Knowledge is not a resource for the intellect. The intellect is a resource for Knowledge. This represents a complete shift, you see, a turning point, as if you have to get on the other side of the mountain within yourself. And that is why it is a journey and has many steps. Knowledge will not be used as a resource for your wandering desires. It just remains silent, and you will think it is not there.

Your intentions have to come from a deeper place within you, and you have to build over time the trust that Knowledge is there, that it is powerful and wise. God is not asking you to believe this at the outset, but you must begin to build the connection to Knowledge and see that Knowledge holds your well-being, your success and your real promise, regardless of the circumstances on the outside. Those are just problems to solve.

The beginning of your journey to Knowledge will require a deep evaluation of where you are, what you are doing, who you are with, your strengths, your weaknesses, your resources, your liabilities, everything. If you are older, you will have to do this re-evaluation in a very great manner because where your life is meant to go and where it is today are not the same. So you have to reconsider everything that you are doing and consider how you got there. The teaching in The Way of Knowledge will show you how to do this evaluation in such a way that will give you wisdom and insight and not simply lead to further confusion and alienation from yourself.

Who and what can you trust? There are really only two choices here in reality. There is Knowledge, and there is everything else—Knowledge within you, Knowledge within others. The choices are simple but require a very deep discernment to recognize on a daily basis. It is this discernment you must cultivate and you will cultivate as you see the contrast between Knowledge within yourself and all the gyrations of your personal mind.

WHO CAN YOU TRUST?

Then you will see what is really trustworthy. Then you will see what must serve. You will see the power. You will see that the mind and the body are vehicles for Knowledge to express itself in the world. And you will see that Knowledge represents your true Self, your deeper nature and the deeper current of your life.

As revealed to Marshall Vian Summers, October 9, 2008

CHAPTER 8

How Will You Know What to Do?

Many people are concerned about the future. As times become more turbulent and discordant, more and more people will be feeling anxiety about what may happen next. Their anxiety will increase in time, for in facing the Great Waves of change that are coming to the world, events will accelerate and economies will become more turbulent and unstable. There will be greater environmental problems, and people everywhere will have to face the reality of resource depletion in the world.

How will you know what to do in the face of great change? Where will you turn? Who will you consult with? Who do you think will lead you into the difficult and uncertain times ahead?

The human family has many great strengths, but one of its weaknesses is it has not planned for the future adequately. People do not look ahead. They do not look and see what is coming over the horizon. And so they find themselves unprepared and overwhelmed by events in life, which could, in most cases, have been foreseen. They had their backs to the future, and the future arrived.

All of a sudden they find themselves underwater or overwhelmed—without work, even without a home. And yet you must ask: "Why did they not see this coming? The signs were there." There are always signs before great events. Sometimes signs occur years before great events. But there are very few people who can recognize these signs and respond appropriately before the event arrives.

People are consumed in the moment and always reflecting to the past. Their thoughts about the future are mostly hopes and fears. People are not trained as children to look into the future with objectivity and to reflect using a deeper Intelligence that God has placed within each person, an Intelligence called Knowledge.

This failure to look ahead and to live life upon self-assuring assumptions represents one of humanity's real weaknesses at this point—a weakness that will lay them vulnerable to immense difficulty, both now and in the future.

Knowledge within you will give you signs and will respond to signs in the world that are significant. For a sign is not merely a message; it is also a teaching. A sign is there not only to warn you, but also to instruct you. Most of the instruction will come from Knowledge within yourself, but sometimes signs actually tell you what to do.

In a very obvious sense, if there is smoke in the building, you know there is a fire. That is a sign, so you respond. But most of the signs that deal with future events are very subtle. They are subtle, but they are also frequent. If you are paying attention, you can see them. And Knowledge, the deeper Intelligence within you, will advise you what to do, will urge you to do certain things and not others.

Every day people are committing themselves to various things, many of which will not work. But because they are not reviewing their decisions with Knowledge, they go ahead upon the urging of others—upon the urging of their family or friends, or urged by their own fear of loneliness, or their own desire for wealth.

HOW WILL YOU KNOW WHAT TO DO?

Every disastrous decision began as a good idea, and people were prompted by some desire or need. But disasters occur because people are not responding to Knowledge and are not referring to Knowledge when they are making important decisions.

You can fool the intellect. It is not difficult to do that. But to fool Knowledge is very difficult. The only way people can be fooled here is that they are not sufficiently aware of Knowledge. They do not use Knowledge. They do not consult with Knowledge. They do not reflect with Knowledge. And so their intellects are easily seduced, easily persuaded, easily assured, easily discouraged.

So We have two problems: people do not look ahead sufficiently and people do not utilize the great gift of Knowledge that God has placed within them.

By looking ahead, you want to look ahead many years, without assuming that everything around you will be the same—without assuming that your economy will be strong or even functional; without assuming that your health care system will be functional; without assuming that you will have all of the advantages that you have.

You must look without assumptions to the very best of your ability. Not with hope, not with fear, but just looking—the way that a sailor would look from the top of the mast of a sailing ship, looking out, scanning the horizon, seeing what can be seen. Not hoping, not wishing, not fearing—just looking. A lookout.

This is a fundamental skill. The beasts in the field are doing this. The birds in the air are doing this. But people are preoccupied with their thoughts, with their plans, with their goals, with their problems, with their resentments, with their issues with one another.

PREPARING FOR THE GREAT WAVES OF CHANGE

They are not looking. They are not listening. So they do not see the signs of the world. And they do not hear the signs from Knowledge within themselves, so preoccupied they are with their own thoughts and feelings.

The future is not as mysterious as you think if you can pay attention, so long as you do not try to be specific regarding dates and times. That always fools people. People will say: "Oh, this date is going to be a very significant year. It is prophesied in the ancient texts." But never hold to a date.

Time is a human creation. The future is always shifting, depending on what is happening today and tomorrow and the days to come. An event may be inevitable, but you cannot tell precisely when it will happen. Therefore, do not prophesy using dates and times. That will get you into trouble, and you will find yourself mistaken and discouraged as a result.

What is important is to see the movement of things—where your economy is going, where the world is going, where your life is going, where other people's lives are going, seeing the movement of things, discerning the direction, discerning the possible outcomes. As long as you do not affix a date to it, it could be very insightful.

People do not ask the really important questions when they are establishing their relationships with one another, or when they are making fundamental decisions because they are not paying attention, and they are not looking ahead.

For example, regarding your work and career, particularly if you are a young person or a person who is changing your work and career, you must seek work that can survive very difficult economic times. You must provide goods or services that will be fundamentally necessary

HOW WILL YOU KNOW WHAT TO DO?

under very stressful economic conditions. No one is going to need other kinds of services. Very few people will even be able to afford them, and so they will largely disappear.

But if you did not have this understanding of the future, then you could make an extremely unwise decision and invest yourself greatly in your endeavor, only to have the circumstances turn against you. And all of this could have been foreseen in the past had you been looking carefully.

To be driven by desires and wishes and preferences is blind and foolish if you do not understand the environment in which they will be expressed. Environments: the physical environment, the economic environment, the political environment, the circumstances; if the circumstances are not conducive, then even the best idea in the world will not function, will not be able to be successful.

People hope the environment will be there for them. They expect it to be there for them. And so they launch themselves on some great economic endeavor or personal endeavor only to find out that their timing was all off, that they did not consider their environment, they did not look far enough ahead. They assumed the future would be like the past, and so they launched themselves at the wrong time and then headed into stormy waters.

Any endeavor that you seek to initiate that will be long term, you must consider the environment very carefully, very objectively, even fearlessly. Will that endeavor have support? Will the economy support this in the future? Will people need this in the future? Will it be a priority for them under difficult economic times?

People are looking backwards, but moving forwards. And so when they crash into things, they are shocked. They did not see it coming because they were not looking, because they were assuming and believing, hoping and wishing while life was moving.

It is like being in an automobile, moving. Everything is coming at you. Would you be looking out the rear view mirror? Would you have your attention diverted for more than a moment when everything is coming at you in your pathway? Life is like this, though more in slow motion. If you are not paying attention, you will have a disaster.

Before great events happen, there are signs. There is a buildup. Even if it is an earthquake, and it cannot be predicted, within yourself you will feel signs. You will feel anxiety, unreasonable anxiety. You will feel caution and fear without being able to identify its cause. You will feel unsettled. And if that occurs, you must set yourself down and say to yourself: "What is the source of these feelings? Are these feelings associated with anything I am doing at the moment? Are these feelings associated with the environment around me, any change in that environment? Is there something I must see or do at this time?"

You must take advantage of these moments of discomfort. They are important. People say, "Oh, I felt bad last night. I was so anxious, and I was disturbed. I could not sleep, but I feel better now."

This is stupid. You do not feel unreasonably anxious for no reason. Something is trying to communicate to you. You are responding to something, perhaps beyond your immediate circumstances, and you are just treating it as if it is just an emotional episode.

Some people think that any fear or anxiety is a bad thing, and so they try to be happy all day long. That is so foolish. They are missing all the signs. They are not allowing their delicate inner guidance system to work for them. They think that any troubled thought or any anxiety somehow runs counter to their expectations and beliefs. If that is

HOW WILL YOU KNOW WHAT TO DO?

the case for you, you should change your expectations and beliefs because you are not taking advantage of the power and the presence of Knowledge to alert you, to warn you, to get your attention, to re-engage you with your environment.

Young people emerge from their schooling, thinking: "Oh, what do I want to do with my life? What would make me happy? What would be rewarding?" But they rarely look at the environment or consider the future. They just think if they do what they want, everything will work out. Well, everything will work out, but not to their advantage or to their expectation.

The environment is very important. If you cease to consider it, it will work against you, and it will defeat you. You have a dynamic relationship with your environment—with the natural environment, with the economic environment, with the political environment, with the social environment, who you are with and their influences upon you. All of these environments can either work for you or against you depending on how you engage yourself with them, and what you choose to do, and how you plan your activities, and the degree of attention you bring to your circumstances.

You should think of your life like sailing across a great ocean, a great ocean which can be calm and peaceful or turbulent and stormy, where the winds vary, where the currents vary. You must have provisions for this journey. You must have skill to take this journey.

You must know what to avoid and how to respond to certain kinds of difficulties. You must always be watching the environment. In a sailing ship of any size out on the ocean, you must be very, very observant and be able to read the signs of your changing environment—the change in the wind, the degree of turbulence in the ocean, reading the clouds.

PREPARING FOR THE GREAT WAVES OF CHANGE

So many people just launch their lives and find themselves shipwrecked somewhere down the coast, floundering out in open waters. Some people sink altogether. They underestimated the power and the potency of their environment. So many business endeavors have failed, marriages have failed, because people did not respect the environment and the demands of life, were ill prepared, did not choose correctly or did not time their activities appropriately.

They were not with Knowledge at the point of decision—not paying attention to the signs, not exerting enough care and caution here, urged on by others, urged on by their own desires, urged on by the fear of losing out, missing an opportunity. They took the plunge only to find that the water in the pool was very shallow.

Humanity is now facing Great Waves of change, greater than anything the human family as a whole has ever had to face before—environmental degradation; diminishing resources; violent weather; economic and political instability; the growing risk of competition, conflict and war over the remaining resources; and a growing population of people in the world drinking from a slowly shrinking well.

That represents a very challenging and difficult environment. That requires great care and skill. How are you going to navigate the difficult times ahead? How are you going to know what to do in the face of change you did not anticipate, change that very few people anticipated?

Better to be cautious here than overly optimistic. But you still must take this journey. You cannot go hide under a rock somewhere or lock yourself in a closet.

HOW WILL YOU KNOW WHAT TO DO?

Do not lose heart, for you have come into the world for a greater purpose, to live in these turbulent and uncertain times. God has placed Knowledge within you to guide you and to protect you and to enable you to take the steps to discovering your greater work and service in the world. But this requires a kind of strength and sobriety, clarity and self-honesty, that very few people have forged within themselves appropriately.

You cannot be fooling around in the face of the Great Waves of change. You cannot be playing on the beach when the Great Waves come, or have your back to the ocean. This is symbolic of what humanity is doing right now.

There are a few people who are aware of the Great Waves of change, who are trying to sound an alarm, but their voices are drowned out by ignorance and indifference and the insistence upon the preservation of affluence. They are alone in their admonitions because people are not paying attention, and they do not have yet the strength to face great uncertainty even though God has placed this strength within them.

The power and the presence of Knowledge is here to guide you and to protect you, but you must take the Steps to Knowledge and become strong in Knowledge. Here you must set aside your fears and your hopes, your nightmares and your strong preferences, to look and to see carefully, to be as objective as you can, to take your environment very seriously, to observe other people to learn from their mistakes and from their strengths, to stop your endless complaining and to end condemnation of others so that you may learn from their example.

Here the whole world can teach you what works and what does not work, what will work for you and what will not work for you. But you cannot learn any of this if you judge the world for disappointing your expectations, if you are consumed with your own interests and desires and problems because you are not paying attention. You are not being a student of life.

So you will not know what to do in the difficult times ahead. You will be swept along. And when you finally decide to take action, it will be too late because everyone will be taking action, and it will be a panic. It will be chaos. The stores will be empty. The banks will be closed. There will be social unrest. And you will be caught in the middle of a situation that you could have escaped had you planned previously.

You must now be very cautious. Caution is different from fear. Fear is a kind of blind reaction, a kind of paralyzing experience. But caution is observant. It is objective. It is inquiring. You are looking to see what you must see. You are taking great care with your actions and your thoughts and your decisions. That is being cautious.

That is what is appropriate here. To be fearful is to cease to be observant. It is to be taken up in your own reaction. It is to be held in place. It is to withdraw and contract. There is little wisdom here.

But if you are not prepared for great change, you will react in this way. You will be paralyzed with fear. You will be outraged. You will be confused. You will have no plan. You will not know what to do. You will blame other people. You will take desperate actions. You will follow other people who are taking desperate actions. You will put your life at greater risk and greater danger.

The Great Waves of change will either defeat humanity or require that humanity grow up and become mature and responsible and united. Humanity must cooperate to face the Great Waves of change. You cannot be a warring set of tribes and cultures, or the Great Waves will just destroy what you have created.

HOW WILL YOU KNOW WHAT TO DO?

This is a great time for humanity. The Great Waves of change have largely been produced by humanity's abuse and overuse of the world; its reckless use of resources; and its lack of concern for the future, lack of awareness of the future and lack of preparation for the future. It is this desperate, irresponsible, reckless behavior that is now getting you into such trouble and will only produce greater difficulties now as you proceed.

Everyone wants economic growth, but you are facing a world in decline. Everyone wants to have as much wealth as they have now and more, but humanity is drinking from a shrinking well. What you want and where the world is going can be so very different here. What you expect and what is coming over the horizon can be so very different.

Do not underestimate the power of nature and the consequences of altering nature. Everyone has had a part in doing this, and everyone will have to bring skill and ability and the willingness to unite and cooperate to face the Great Waves of change that are coming to the world.

Knowledge will teach you what to do step by step, but you must undertake a real and deep evaluation of your life and current circumstances.

What is the sustainability of your work? Of your home? Of your transportation? Where do you get your food and energy? What are the resources of your community? How will you get to work if there is no petroleum available? How will you deal with food shortages and ever-increasing prices of everything?

You have to think ahead now. It takes courage. It is unsettling. It might even be terrifying at first. But if you keep your attention here, you will gain a greater objectivity, and the waves of fear will pass. And you will learn to become observant and cautious, careful in what you do.

But it must begin with a great evaluation of where you live, how you live, how you travel, how you use resources, the strength of your relationships. Do the people around you have any sense of what is coming over the horizon? Or are they just foolishly playing in the sand or carrying out their own objectives without any awareness of the great change in your environment?

This evaluation will take time because it requires many things, and it will require changing many things and redirecting perhaps the course of your life entirely. But time is the problem. Time is of the essence. You do not have a great deal of time to prepare for the Great Waves of change. Every month and every year are very significant now.

If you are living in the wrong place, if you are poorly positioned, you will reach a point when you cannot do anything about it except take a great loss. You thought the future was going to be like the past, so you invested in that way, under that assumption, and now you find yourself in jeopardy.

This is the challenge of life. It has always been the challenge of life. But now the challenge is so much greater.

The people in wealthy nations have lost contact with nature. They have lost their ability to be discerning and cautious and aware. They have become complacent and self-assured, living on assumptions that their wealth would continue, and their privileges would be supported, and their environment would continue to provide endlessly for their desires and their needs.

HOW WILL YOU KNOW WHAT TO DO?

While the poorer people are facing the reality every day—the problem of resources; the problem of security; the problem of protecting themselves and their families; the problem of acquiring sufficient food, water and medicine—they are wiser by far than the wealthiest amongst you. They are closer to reality. And now reality will reassert itself everywhere, even in the rich nations.

No one will escape the Great Waves of change. And the wealthier you are, the more you have to lose, and the more concern you will have in protecting what you have, and the greater will be your vulnerability in facing a world of needy people.

This is a problem that will either defeat humanity or unite and uplift humanity. The decision is not only with governments or large institutions. It is with every single person. Every person must choose whether they will struggle, compete and fight or whether they will find ways to sustain themselves in concert with others.

It is a fundamental decision in the face of grave circumstances. It is life. You may think it is calamitous, but you are really returning to a more authentic, grounded and genuine life. Instead of playing with your little technological toys, instead of losing yourself in your romances and your hobbies, you have to come back to life, to stand at the edge of life with all of its uncertainties, opportunities and dangers. Do not think that everything will turn out appropriately, for everything will turn out, but it will be a disaster for so many people.

You have time to reconsider your life, to gain a greater degree of wisdom and discernment here, but you do not have much time. That is why this Message is so important. That is why the reality of Knowledge is so important. That is why looking far into the future with clear eyes is so important. That is why making plans and preparations and discerning how you will be able to take care of other people are so important.

There will be people beyond your family that you might need to care for—the elderly, children without parents, disabled people, or simply people who just fell apart in the face of the Great Waves of change. You may have to take care of some of these people to a certain degree. How will you do that and with what?

These are things you should be thinking about now, before the storms arise in great number. You should be building your ark before the rains come and not wait until the last moment when there will be nothing to use to build an ark.

This is why there is a New Message from God in the world because humanity is unprepared for the consequences of its own actions in the world—its lack of preparation, its lack of wisdom, its lack of responsibility and accountability—the results of which are now mounting and will continue to mount, relentlessly, as you move forward.

God loves humanity and does not want to see human civilization— which has taken so long to build, which has such great promise and great qualities—God does not want to see this fail and collapse. God does not want to see humanity fall prey to forces from beyond the world who are here to take advantage of a weak and divided humanity.

That is why there is a New Message from God in the world. And that is why it is here to warn you about the Great Waves of change. It is here to provide the Steps to Knowledge and the preparation that will be needed—the preparation within each person, in how they engage with themselves, with other people and with the world itself.

HOW WILL YOU KNOW WHAT TO DO?

This will require great strength and courage. But your life is on the line, and so there should be sufficient motivation for you to have this strength and courage. And you must be very compassionate because people around you will become ever more distraught and confused. They will engage in self-destructive behaviors. They will act foolishly. And they will be delusional.

You must have great compassion, for humanity as a whole is not prepared for the Great Waves of change. It is not prepared for the Intervention that is taking place in the world by races from beyond the world. It is not prepared for life, for eventualities.

You must see the truth of what is coming, and you must have the strength, the power and the motivation to act—appropriately, wisely, carefully—using the wisdom of others to help you, using the resources of your community, reaching out to other people.

For it is not about stockpiling food for the future. That will only work very temporarily. You are facing a very long set of circumstances. You have to reset your life, reposition yourself.

That is why the Great Waves of change, though calamitous and extremely dangerous, also can be redemptive. They require a greater honesty, a greater capacity and a greater responsibility in people. This is redemptive because you have come to the world to serve the world. If that service is not being activated, if it is not being realized and expressed, then you are failing your fundamental mission in life.

Whether you are rich and live in splendor, you will feel uncomfortable, restless and anxious because you are not engaged in your greater activity in the world. You are not fulfilling the need of the soul. You are not finding and expressing your greater purpose for being here.

PREPARING FOR THE GREAT WAVES OF CHANGE

Once you get beyond your fear and anxiety, you begin to see that the Great Waves of change are connected to who you are and why you are here. And that is a very significant realization. That presents a real shift in how you will see and understand the greatness of your times and the power that God has placed within you—to face them and to serve them and to live meaningfully within them.

As revealed to Marshall Vian Summers, August 19, 2008

CHAPTER 9

Realizing the Need to Prepare

It is very important that people realize their preparation for the future is essential. It is very important that people realize that the future that they will be facing will be very calamitous and difficult.

While there is much speculation about what is occurring in the world today and many projections regarding the future, there are actually very few people in the world who can see fully what is coming over the horizon. That is why God has sent a New Message into the world: to prepare humanity for a converging set of changes and difficulties unlike anything humanity has ever had to face before.

Indeed, the changes that are coming will change the living conditions of humanity. And these powerful converging forces will either defeat you or uplift you, depending upon how you see them and the degree to which you prepare for them.

The Great Waves of change are coming to the world. They are already striking the world in many places: changing climate, violent weather, the depletion of essential resources, political and economic instability and a growing risk of conflict and war. These will set in train forces of change that are largely unanticipated, even by the leaders of nations, even by your leading scientists.

PREPARING FOR THE GREAT WAVES OF CHANGE

For you might consider the impact of, let Us say, political and economic instability because of corruption, incompetence and mismanagement of financial systems. You could see that that alone could produce havoc and a great loss of wealth for many, many people, even in the wealthy nations. But when you combine that, which is already in motion now, with the other Great Waves of change and the consequences that they would produce, both singly and collectively, then you are dealing with a very complex situation.

While people may not see what is coming over the horizon, their feelings are giving them signs: a growing sense of anxiety, a growing sense of anticipation, feeling out of control regarding their circumstances and their resources.

Surely, the poor people of the world are being impacted directly, facing ever-greater costs for everything that they need. And they are facing the impact of violent weather destroying crops, devastating towns and cities, creating immense instability and a human welfare problem. They are facing the growing risk of disease, even pandemic illness. They are already paying the price for the indulgences of the rich, for the lavish lifestyles and the overuse and misuse of the world's resources.

Surely, they can tell the story of the Great Waves of change. But it is only the beginning, for the Great Waves are immense and continuous. And if you have the courage and the objectivity to face this and to begin to discern the Great Waves of change and the impacts they are already having around the world, then you must ask yourself honestly, "How will I know what to do in the face of great change?"

You may lay out plans. You may have schemes. You may even stockpile food. You may try to live defensively. But this is not an

REALIZING THE NEED TO PREPARE

answer. You cannot stockpile food for a lifetime. Indeed, all your plans for defense may prove to endanger you even further.

For example, some people think that they should move out into the country and grow their own food, having no idea, of course, what this would really require and how arduous this would be.

If they followed such plans, they would put themselves in ever-greater danger and vulnerability. Unable really to provision themselves, they would now be far away from sources of food, energy and security. So even amongst the very few who sense great change and upheaval are coming, often their plans are counterproductive and even self-destructive.

In most cases, people are trying to change their lives in ways that they have always wanted to change their lives, thinking, "Oh, I have always wanted to live in the country," or "I have always wanted to go and live in this particular place, and now I have an excuse to do that." But this is not really responding to the great change that is at hand and that will continue into the future. This is merely one indulging in their desires and their fantasies.

It is unfortunate that so many people do not have the courage and the strength within themselves to face the Great Waves of change. They simply will feel overwhelmed and therefore will go into denial, thinking, "This cannot be. This cannot be true. I will not look at this." And they do this at their own peril, of course.

For if you do not look, you do not see. You will not prepare. You will not make plans. You will not put yourself in a more secure and stable position. You will just keep living life the way you have been living it until something happens to you.

This will be the unfortunate fate of many people. It is not because nature and life are so surprising, for the Great Waves have been building for a long time. And there has been much evidence that they are coming and that they are occurring. And this evidence has been growing.

The signs are all around you. The world is telling you what is happening and what is coming. These signs can inform you if you can face them and read them accurately—without preference, without fear, without trying to turn the situation into something you want, without trying to fulfill unfulfilled desires or fantasies.

The change at hand is so immense that many people simply are not strong enough to face it. They would rather turn away and put their backs to the world, thinking and believing that someone else—science or government or even God—will simply take care of all of this for them, and that they are going to live for the moment.

They do not want to be afraid. They do not want to live in fear. This is what they think. But really what they are saying is: "I really don't have the courage to face this. It is too much for me."

So while the world is giving them signs, and the deeper Knowledge that God has placed within them, at the level of Knowledge, is giving them signs, and their feelings are giving them signs, they choose not to look. They choose to pretend. They think it is a problem for the poor people in poor countries, or a problem for other people, or a problem for governments or for economic systems to solve. Even the very well educated, even those who consider themselves to be sophisticated will succumb to the same fear, thinking it is something else.

REALIZING THE NEED TO PREPARE

There will be many excuses, and the signs will be dismissed, and people will think they are being very intelligent because they think the future is going to be like the past. They will think to themselves that they have been through this before. It is just an economic cycle. The climate may be changing, but it will not change for a hundred years. They will think there is plenty of food production in the world. There is plenty of petroleum and natural gas. They will think that technology will bridge the gap in all cases.

They do not realize that their minds are governed by fear and weakness. They think they are being reasonable and logical, but they do not realize that they are facing a future that will be unlike the past.

They are not asking the right questions: What is coming? Why is it coming? What does it mean? And how do I prepare? Those four questions are key questions. Many people ask, "Well, who is going to take care of this problem?" Or they think it is another nation that is using too much energy. Or they think that it is business and commerce that are evil and destroying the world when they themselves are the beneficiaries of the exploitation of this place.

People are not seeing, so they are not asking the right questions. If you do not ask the right questions, you do not have useful answers.

Yet there are a few people who are sufficiently uncomfortable that they will begin to look and try to discern what is happening. They realize that pretending or that speculation or that projecting one's understanding from the past is not adequate or appropriate now and that humanity has not been here before, humanity as a whole, that is to say. They will look and they will realize eventually that even given all the alternatives and all of the solutions for various aspects of the Great Waves, that in fact they do not have an answer for the whole phenomenon.

PREPARING FOR THE GREAT WAVES OF CHANGE

Some of these people will give up. They will simply think it is over; this is the end. And they will fall into depression and prolonged anxiety. Others will just have hope that something good will happen, and it will all work out some way, even though they cannot see how that could possibly happen.

But the critical question here is: How will you know what to do in the face of great change? The question is important because the key word is "how" will you know what to do?

For in truth you are facing a situation that has so many dynamics to it that planning and scheming alone are not sufficient. You will have to determine at many critical points what you are going to do. And you will have to have the freedom to change course in your life—to change your work, to relocate, even to leave others who will not and cannot prepare, if that becomes necessary.

You have to have this freedom and this flexibility to respond to the Great Waves of change, or you will simply stay in place and they will overtake you—robbing you of your employment, robbing you of your home, robbing you perhaps of your ability to function, depriving you of even your most essential needs. That is the power of the Great Waves of change. Do not underestimate the Great Waves, or you will put yourself in immense peril.

Your first great challenge is to face the great challenge, not with plans and schemes or demands for solutions. You must face them. And you must face them to discern their movements. And then you must read the signs within yourself and within the world that will tell you what you need to do.

Are you living in a place that can be sustained into the future? Does it have continuity? Will it be livable five years from now? Ten years

REALIZING THE NEED TO PREPARE

from now? Do not think it will be easy to move or relocate in the last moment, for it may not even be possible.

Is your employment sustainable into the future? The work that you do, will it really be fundamental and needed within the reality of the Great Waves of change? If you are selling non-essentials, non-essential goods and services, it is pretty evident that your career or your work will not have a future. No one will be interested. No one will have the money to spend on these things.

If you are young and educating yourself for the future, you must think of these things very seriously. You must have a career or a job that can survive the most difficult of circumstances. No one is going to be indulging in frills in the future, except perhaps the very rich, and even they will have to hide from others—hiding their wealth, not flaunting it in public.

This is a gift of Love, a gift born of Love and concern for humanity from the Creator of all life. You may feel this is immensely scary and disconcerting and inconvenient, but it is a gift to prepare you so that you may survive the Great Waves of change and so that ultimately you may become a real contributor to humanity.

God has placed Knowledge, a deeper Intelligence, within you to guide you and to protect you. But this Intelligence exists beyond the realm and the reach of your intellect. You cannot figure it out. It alone knows what is coming and how to prepare. And it will prepare you not only to survive, but to contribute for that is why you have come into the world at this time.

For it is true you have come into the world to serve the world in the face of the Great Waves of change. This represents your greater purpose and calling in the world. Your calling here is not to be an artist. It is not to pursue a personal agenda. It is not to try to fulfill yourself in some line of work or to forever seek romance and sensation. These things will prove to be so false in the future. And yet they are so false even at this moment.

You are here to use your talents and to respond to a world in need. It will tell you where you must go, what you must do and where you are to serve and to contribute your unique gifts to the world. This is such a different emphasis, you see, than people trying to fulfill themselves through work, through business, through wealth, through pleasure, through art. It is a total reconsideration of one's life, a deep evaluation.

Some people look at the evidence of the Great Waves, and they start thinking about solutions, but their solutions will be inadequate. It will take more than a few solutions to be able to navigate the difficult times ahead and to contribute to humanity's well-being and future sufficiently.

It will take thousands of solutions, all supported by many people. There is no one solution. This is bigger than your technology. This is bigger than your economic system. This is bigger than your political structure.

Humanity is facing a world in decline—a world of declining resources, a world of environmental change. Do not think your mighty armies can defeat this, or your great and destructive weapons can overcome this.

You must begin to rely upon a deeper Intelligence that God has placed within you. You must go beyond your intellect here for the real guidance you will need. Your intellect can solve immediate problems, small problems, and can deal with the details of what you realize you

REALIZING THE NEED TO PREPARE

may have to do, but the guidance and the clarity must come from a deeper place within you.

Without the guidance of Knowledge, you will follow other people's prescription, if indeed you do anything at all. And many of their plans will be really disastrous and unwise. You will react out of fear and even panic, and your decisions will not be wise, and they will cost you whatever resources you still have, placing you in greater jeopardy.

God knows what is coming. God has sent a New Message into the world, a Message that contains a warning, a blessing and a preparation. The preparation is mysterious because it does not follow your ideas or your notions. There are only parts of it that you can explain to yourself. Some of it will appear very practical; some of it may not appear to be practical based on what you think is practical.

The Mind of God can see a hundred different changes. You can only see one or two. That is why the power and presence of Knowledge within you is your guiding light. If your life has to change many times, it can navigate this, whereas your intellect will simply become confused or enraged or perplexed or feel defeated. Knowledge within you will show you where you must go, whom you must meet, what you must do, in every circumstance.

The preparation will teach you to become still inside—to listen, to discern the voice of Knowledge within yourself and within other people. It will cut through the confusion of ideas to the certainty of reality.

Time here is critical. If you wait, your choices will diminish. If you wait for everyone to agree with you, it will be too late to prepare. You must build your ark before the rains come. If you wait, there will be nothing to build with.

PREPARING FOR THE GREAT WAVES OF CHANGE

You must secure your position while you can, while things seem quiescent, while everything appears to be normal. For if you wait, you will not be able to make these preparations.

Human populations will panic. There will be a stampede. It will be chaotic and very damaging. You must respond to Knowledge and follow Knowledge now. If you wait, it will be chaos.

You may ask: "What am I really preparing for? Is this going to be difficult or terrible?" You must prepare as if it were going to be terrible. And you must prepare with your understanding that you are not merely being guided to protect yourself. You are being guided to a position where you can be of service to others. That is very different than simply taking a defensive position.

You may not be able to preserve all of your wealth, all of your advantages in the face of the Great Waves of change. Preserve what you can, lasting into the future, but Knowledge will guide you to be of service, for that is your greater purpose for coming here at this time.

It is no accident that you are here in the face of the Great Waves of change. Your intellect cannot account for this, but at a deeper level, it is surely known.

What will redeem you is discovering and fulfilling your greater purpose for coming to the world. To do this, you must engage with the world as it really is, not as you want it to be or you think that it should be.

Love follows acceptance. You cannot love the world if you cannot accept the world. And you must accept the world now even in the face of the Great Waves of change if you are to be of service to others and to fulfill your greater destiny here.

REALIZING THE NEED TO PREPARE

At the outset, the Great Waves are immensely shocking, discouraging and disappointing. But if you will abide with them, you will discover over time that there is a very great connection between the Great Waves of change that are coming to the world and who you are and why you are here. But at the outset, you may not be able to see this, for you are in shock and awe. And you are in denial and anxiety—running around, confused, upset, not knowing what to do.

But once you calm down, and once you ask honestly: "What must I do?" then you are in a position to begin your preparation in a sensible way. It is as if you are preparing for a great storm, a storm that will last for decades. You only know how to prepare for a little bit of it, so you must be guided in this matter by Knowledge.

Even the experts really cannot answer the questions for you now. For many of the experts cannot see what is coming over the horizon, and no one has a real plan that will be workable in all situations.

That is why you must be alert and attentive. Look for the signs of climate change. Look for the signs of diminishing resources. Look for the signs of economic and political instability. Watch around the world the escalation in the price of food and fuel. Pay close attention to difficulties regarding drought and water.

Look at these things ongoing to get a sense of the movement of the Great Waves of change. That will indicate to you what you must do in specific matters and how much time you really do have. Time wasted now is time lost. Time wasted through denial and speculation and through blaming others is time lost, shrinking your window of opportunity.

You must come to terms with reality. It is not what you think it is. But your deeper nature can respond to it and is responding to it already.

PREPARING FOR THE GREAT WAVES OF CHANGE

What is coming? Why is it coming? What does it mean and how do you prepare? And the most primary question of all: How will I know what to do in the face of great change?

This aims you in the direction of Knowledge, for your mind does not have an answer. And the answers you think you have will not be sufficient, so you must now become very attentive and gain as much objectivity as you can.

You must become educated about the Great Waves of change in the world. You must stop blaming other people and use that energy to prepare your life and to assist others in preparing their lives. You must prepare for this storm.

If you can, you will be able to navigate the difficult times ahead. And you will be able to be of tremendous service to others—beneficial service, a source of strength and inspiration for others. For there will be so many people who do not have this strength and this inspiration.

A New Message from God has been sent to warn people of the Great Waves of change that are coming to the world and to teach The Way of Knowledge so that people will know what to do, and to teach about relationships at the level of a higher purpose in life.

The New Message is here to prepare you for your encounter with intelligent life from beyond the world and to face an Intervention from certain races who are here to take advantage of humanity in the face of the Great Waves of change.

This is the New Revelation. It holds with it an immense challenge for an unaware and unprepared humanity. It is a great deal to face, but this is why you have come. You have not come simply to be a mass consumer, a locust upon the world. You have not come here to lose

REALIZING THE NEED TO PREPARE

yourself in hobbies and addictions or to work slavishly for some little conveniences in life.

You have a greater purpose and mission here. And the Great Waves of change are the context—the environment in which this mission can be called out of you, realized, experienced and expressed.

This is the answer to your prayers. If you cannot follow the Knowledge that God has placed within you, then do not ask for a miracle. It is unreasonable to do this. There will be some miracles, but you cannot rely upon a miracle in the last hour to provide what you will need and to give you stability and security in the face of the Great Waves of change.

There will be great human suffering in the future. Masses of people will have to relocate. There will be great shortages of food and fuel, even in the wealthy nations. There will be a breakdown in civil order in large cities. And some countries politically will collapse.

It is really what you have been expecting, but it is greater than you have been expecting.

The Great Waves of change are largely the consequence of humanity's misuse and mismanagement of the world's resources. You have not attained stability and security yet in this world. You are still in a very adolescent phase of humanity's overall development. You are discovering your powers and strengths in certain areas such as technology, but you do not have the responsibility or the restraint or the wisdom to create a stable living situation for the human family in this world. And you are grossly unprepared for the reality of a Greater Community of intelligent life in the universe, and what your contact with this Greater Community will mean and require of you.

PREPARING FOR THE GREAT WAVES OF CHANGE

It is because of this lack of awareness and preparation that God has sent a New Message into the world for the protection and advancement of humanity. It is not here to replace the world's religions, but to strengthen them and to unite them and to give them a greater dimension of understanding and application.

For God is the Author of all the world's religions, and all the world's religions have been changed by humanity. What is essential and true within them has become lost in so many cases.

But humanity must find its greater strength, the greater strength of Knowledge. It is not simply the strength of your willpower or your ideas or your social associations. It has to be a deeper strength within you—the kind of strength one faces when going into battle.

It must be clear; it must be objective; it must be determined. But it also must be compassionate and flexible. The Great Waves will not happen according to your prescriptions or expectations. You must continually read the signs to know how to respond—the signs from the world and the signs from Knowledge within yourself.

This is where humanity will finally grow up and outgrow its ceaseless conflicts, its religious divisions, its pathetic indulgences and addictions and its tragic wasteful use of the world.

You are now being forced by necessity, whereas before you were encouraged by reality. Now you will be forced by necessity. The outcome could be positive overall, but it will be very difficult to achieve such an outcome and will require fundamentally a restructuring of the world and of human civilization—not out of precept or philosophy but out of necessity.

REALIZING THE NEED TO PREPARE

People will lose individual freedoms because society must survive. If [society] is guided with wisdom and clarity, people will have fewer options, but in time will gain a greater stability in the world. War will have to come to an end as being self-destructive and self-defeating in the face of the Great Waves of change.

It is the passing of an age, but the passing itself will alter the course and destiny of humanity, depending upon how humanity responds and prepares for the Great Waves of change.

This warning is a blessing. It is a gift of love. It represents the desire for humanity to evolve and to unite and to preserve the world so that it may have a future, and to prepare for its encounter with the Greater Community of intelligent life, where your self-sufficiency in this world will be necessary to secure your own freedom into the future.

You do not yet realize the importance of your life, and you do not yet realize the great change that is at hand and how the two are directly connected to one another. This is a calling for you to respond, to be responsible, to be able to respond, to respond with clarity, objectivity, courage and determination.

You cannot be fooling around in the face of the Great Waves of change. You must avoid denial and all of the attractions that would dissuade you, preoccupy you and distract you from this greater focus. Your life and well-being and the condition of your existence are dependent upon this.

Do not think that because you live in a wealthy nation or that you are perhaps affluent that that will insulate you from the Great Waves of change, for it will not.

PREPARING FOR THE GREAT WAVES OF CHANGE

This is calling you out of the shadows, out of confusion, out of self-obsession, out of fantasy, into a greater participation with the world and a greater connection within yourself and with other people. It is entirely redemptive for you if you can see it clearly.

It will free you from internal division. It will free you from your unhappy past. It will free you from your own frustrations and angers because the future and even this moment will demand so much of you that you will have to give your preparation your full attention.

God heals you by giving you something important to do with your life. You escape the past to live fully in the moment and to prepare for the future. And the future you are preparing for will now require everything from you—your full attention, your complete involvement. It will require you to become connected to Knowledge within yourself and to take the Steps to Knowledge so that you will have a real foundation and a real source of strength in your life.

You have always needed these things, of course, but now you will need them fundamentally. For life will require you to become strong and compassionate. It will call Knowledge out of you. It will call out of you your greater gifts if you can prepare accordingly.

These are the signs you have been waiting for. This is your time, the great time of decision and preparation for humanity.

God's New Message will teach you how to prepare and how to see and know and respond to ever-changing circumstances. It will renew your native abilities, and it will teach you the path to take to re-engage with Knowledge so that Knowledge may be your foundation as other foundations collapse around you.

As revealed to Marshall Vian Summers, September 8, 2008

CHAPTER 10

Seeing What Is Coming

People want to know what is coming for humanity. They want to be reassured. They want to know that their investments in the world are secure. They want to have their expectations confirmed. They seek not to know the truth, but to have their fears allayed, to quell the growing anxiety that they may be feeling about the future of humanity.

They focus on the problems of the moment, for they seem comprehensible and solvable, perhaps, while the greater problems are confounding. They feel weak and impotent in the face of them. They will leave that to the experts, whoever they may be. Or they think that providence will secure their future because they are true believers.

Very few people want to tread far beyond their immediate circumstances, far beyond the events of today and tomorrow and the week and the months to come. They have their plans, yes, but they do not want to see, though God has given them the eyes to see and the ears to hear, as God has given the birds the eyes to see and the ears to hear. The birds are watching and listening, and the animals in the field are watching and listening. But the intelligent creatures, the creatures of conscience, well, they are ambivalent, or they avoid this altogether.

PREPARING FOR THE GREAT WAVES OF CHANGE

God is sending the signs. The world is demonstrating its condition and its direction. But who can see it? Who can respond? Who has the courage and the self-confidence to face uncertainty on a larger scale without answers, without reassurances, without confirmation of their preferences, without a sure and secure idea that their investment in the past will prove to be effective in the future?

People want many things, but that is, for the most part, to secure their previous investment or to gain more of what they want. Who wants to see and know and face the uncertainty of death? Better, they think, to rely upon the assurances they give themselves and the assurances the leaders of the nation provide. It is a fool's paradise. It is a false hope. And the calamity that comes upon people seems sudden and unexpected.

Yes, there were indications of a problem. Yes, there were signs that they will reflect upon in retrospect. But they did not see it coming. They did not see the great change in their health, the great change in their economy, the great change in their working conditions or workplace, the great change in their relationship, the great change in their community. They did not see it coming.

So they suffer not only the consequences, but the shock of the consequences. And they are angry, and they are frustrated, and they complain about the government or the leaders or the rich people or whoever seems to be responsible for their plight and dilemma, when in reality it was a failure to see and a failure to respond.

The birds in the air, the beasts in the field, they have no assurances, so they are always paying attention to their environment. They do not have a mind that tells them that everything will be okay and they really do not need to be so observant and so vigilant. They do not have a mind to tell them that someone else is going to take care of

SEEING WHAT IS COMING

this problem for them. They do not have a mind that haunts them with fear—the fear of future loss, the fear of not having, the fear of rejection, the fear of injury, the fear of death.

They do not have the burden of foreknowing. They do not have an awareness of the future and the past. They are not living in regret, trying to compensate for the past. Their lives are simple and basic, but they are demonstrating something that the more intelligent creatures must exercise and regain.

For most people in the world have lost the vision and the awareness and the vigilance that are necessary to be in life successfully and to be able to gain a greater self-confidence and the assurance that comes with realizing that you have come from beyond the world and that your journey here is temporary, only to prepare you for the next step beyond.

Your assurance is not ignorance. Your assurance is not simplemindedness. Your assurance is [not] a lack of intelligence. Your assurance is a greater strength and a greater power that the Creator has given you—a deeper Mind, a permanent Mind, a fearless Mind, the Mind of Knowledge.

Your awareness of the future and the past makes you vulnerable to tremendous fear and anxiety and to the burden of regret and sorrow, grief and self-recrimination. It is as if your mind is a tyrant and a terrible burden all at once. But beneath the surface of this social mind, this worldly mind, is the deeper Mind of Knowledge.

If you are to be aware and intelligent, if you are to be able to plan for the future and create wonderful things and alter your life and alter the landscape of the world, beneficially, you must have this greater power to guide you. Or you will be too afraid to see, too afraid to know, thinking it will only add to your burden of fear and anxiety, fearing that it will only add to your apprehension and sense of helplessness and hopelessness that lurks beneath the surface of all of your self-confirming and self-assuring ideas.

Even those who are optimistic are still driven by fear. Their optimism becomes a replacement for seeing, hearing and knowing, for being aware of one's condition and one's environment and what is coming over the horizon. To feel better in the moment, they choose to be blind and self-assuring and to only listen to those things that confirm their self-assurance.

Temporarily, they seem to find reprieve and are not as miserable as those who have a greater awareness, or so it seems. But their predicament is the same and is compounded by the fact that they really think that their attitude is going to change the circumstances of their life.

The greater trust must come from a deeper place. This produces a confidence, a strength and a determination that is so much greater than belief or self-control or insisting that the world be the way you want it to be or expect it to be—living in denial, living in delusion, living in fear.

Intelligence and awareness are a great burden until you can gain the power of Knowledge and allow it to direct you and to give you the eyes to see and the ears to hear, without the constant burden of fear and anxiety.

Human civilization is at the brink. You are facing a world in decline—a world of diminishing resources and growing population, a world of ever-increasing political and economic instability, a world

SEEING WHAT IS COMING

where the basic resources of life will become endangered and more difficult to acquire. But who is paying attention to this?

The average citizen wants to be self-assured because they are not strong with Knowledge. The average citizen wants to involve themselves in their hobbies and their pleasures and pastimes and not think about the changing circumstances of the world that are going to determine their future and their decisions, because they are not strong with Knowledge.

Even the leaders of commerce and government and religion are suffering from the same incapacity. They think that the problems of the day will resolve by some kind of magical event—by the forces of the marketplace, or by political ideology—as if a waving of a wand is going to alter the course and the direction of their nation and their peoples.

And if they are really aware, if Knowledge is strong, who can they communicate with? The people that they serve, the nation that they serve, do not want to hear these things. They are too afraid. They are too weak. They are unprepared. They just want to believe and be reassured. They want someone else to take care of the problem for them.

So even those in leadership who are strong with Knowledge and can see and respond to the changing circumstances of the world, they are so hindered and isolated. Everything they communicate now has to be conditioned and limited to the capacity of those who will respond to them.

How can you blame the leaders when the population wants to be blind and foolish? And those who are not struggling to survive and hold onto what they have, they are lost in their hobbies, their interests, their romances, the self-obsession over their health, or furnishing their dwelling place, or whatever it might be. They do not see the next great wave of change that is coming, the next economic tsunami. No, they are living on the beach, listening to the radio, enjoying the warmth of the sun.

You who have the opportunity to hear and to prepare and to respond with God's New Revelation, you have the opportunity to gain the vision, to gain the awareness, to gain the strength, to gain the power of Knowledge. But you must step aside from the culture, for it is not responding; it is not preparing; it is not gaining vision. It is lost—a great tragedy, the misuse of so much intelligence, skill and talent, wasted and lost because people do not have the strength and the courage to respond.

People think a new political party, a different ideology, is going to make things different, but they are not solving the problem at the level at which it exists. This will only delay the time in which humanity can prepare. You see this all around you.

People will think, and they might even say, "I don't want to be aware of these things. I want to be happy. I don't want to be bothered with these things. Everything will work out fine."

So the man or woman of Knowledge moves to higher ground before the evidence of calamity. They are not standing on the beach, collecting the shells as the ocean withdraws, preparing for the great wave that will devastate everything in sight.

The animals move to higher ground. The birds withdraw. And the people are asleep, sitting on their veranda, watching the ocean recede, wondering what that means.

SEEING WHAT IS COMING

What can God do for the person who will not see, who will not respond, who will not pay attention to their environment? Pray to God for peace and equanimity, for advantages or to avoid difficulties, for healing, for economic opportunity, but what can God do for the person who will not see and will not respond?

What God has to offer, they do not want. What God wants to show them, they do not want to receive. What the signs of the world are alerting them to, they do not want to pay attention to these things.

So the tragedy is set in motion, you see. What can you do but prepare yourself and prepare others who are willing to respond and bring them the New Revelation, which holds the warning and the blessing and the preparation itself?

You cannot change their minds. And if they cannot respond to the Revelation, what can you do for them except invest yourself endlessly in trying to point things out when in fact they are choosing not to see.

If God cannot reach them, how will you reach them? If the Revelation cannot prepare them to secure and protect their lives, what can you do?

And then there are so many people who are hungry, who are homeless, who have nothing, who are desperate and destitute. You must always serve them. They are not in a position to respond to the world. They are struggling to survive, and this accounts for so many people in the world.

When We speak of those who do not respond or will not respond, We speak of those who have the ability to respond, who have the resources to prepare. They are the ones who are failing the calling of the world. It is not the poor and the destitute. Their objective is to meet the requirements of the day, the basic requirements, to try to survive under political or religious oppression.

No, We speak to those who are free or who have greater freedoms, who have greater wealth and resources. They are the ones who are failing humanity, and failing themselves.

You pray to God for deliverance, but God has given you the eyes to see and the ears to hear and the deeper Knowledge to guide you. But if you do not use these, what can God do for you?

God is not managing the weather. God is not managing the economy. God is not managing your health. God is not managing your relationships. To think God is managing these things is to think of God as a kind of errand boy for your personal wishes and your wandering desires. It is pathetic. It is ignorant. And it is arrogant.

If human civilization fails, it will be because those who were able to respond failed to do so. And it is a tragedy for those few who are responding, for they look around and they say, "Well, who is with me in this?" and they find that few of their friends and family really want to be aware of the Great Waves of change that are coming to the world or humanity's position in the universe, facing Intervention from races from beyond. Who wants to know of this?

The governments of the world and the religious leaders in particular must be strengthening their people and preparing them for great change instead of telling them that God is just going to take care of everything if they believe. God has already provided you everything you need, but if you do not use what has been provided, how can you ask for more?

SEEING WHAT IS COMING

God has given you the power, the gift, the skill, the intelligence. God has given you a fearless mind to guide you and has provided in the New Revelation the Steps to Knowledge so that you may discover this and experience this for yourself.

Do not complain. Take all the complaints and turn it into preparation. You must build your ark, and you must encourage others who can respond. For everyone who cannot respond, you pray for their well-being. You cannot assure their future.

In nature, it is survival of the fittest. In the Greater Community, it is survival of the wisest.

People want many things. Many pray for these things. But the prayer really must be to give you the strength to see and the power to know and the courage to face your circumstances objectively and the strength to change what must be changed and to secure what must be secured. That is the prayer that engages you with the gift of providence.

God cannot and will not govern your life individually. The Lord of all the universes is present but is not manipulating your circumstances.

You must accept that many people will fail in the future in the face of the Great Waves of change. It is a great tragedy and is unnecessary. It was not meant to be this way. But what God wills and what people choose are not the same. What God emphasizes and what people prioritize for themselves are not the same. That is the dilemma. It is a tragedy throughout the universe, not limited to the human family.

That is why in the Revelation the teaching about Knowledge is so fundamental. It was always in the world's religions but has never been emphasized sufficiently. If that is what God has really given

PREPARING FOR THE GREAT WAVES OF CHANGE

you to navigate the difficult and unpredictable world, then that is the most important resource you have. That is the greatest strength you have. That is what will give you certainty and integrity, honesty and strength.

Neglect this and you will rely upon the assumptions of others and the appearance of things and the capricious nature of human awareness and determination—all things that can change in a moment, all things that do not prepare you for the future or enable you to strengthen your life, your relationships and your endeavors.

People want many things, but it is not Knowledge that they want. It is not awareness that they want. And if they had awareness, they only want pleasurable awareness. They want to be happy, to be ecstatic, to immerse themselves in spiritual euphoria perhaps or believe that God is just going to take care of them because they are very devout people.

But the failure to see, the failure to know and the failure to prepare are all the central problem. And religion, in its purest form, in its true intent, is here to meet this problem.

How many people have perished throughout human history because they were not paying attention, because they did not know what was approaching them, because they were not strong enough to face the realities of life? Their numbers are countless.

Look at the people around you and ask yourself, "Is this person prepared for the Great Waves of change?" and "How would this person respond or be in the face of such great and mounting uncertainty and upheaval?"

Intelligence in the universe is the will and the desire to adapt and to respond. It is not simply cleverness and creating gadgets or being

SEEING WHAT IS COMING

witty or being creative in the arts. Why would the birds survive the tsunami and not the people on the beach? Why would the elephants head to higher ground while the people are sitting on the veranda at the hotel? What is intelligence? Who is intelligent? Who can respond? Who is able to respond? Who is responsible?

These things We say will save your life and the lives of those you love. They are a gift of tremendous love and compassion, but they must be given strongly. For people are asleep and weak and self-indulgent. They are not in tune with the world around them. They are not in tune with their own Knowledge, their deeper nature. They are not in tune with what God is revealing and what humanity must see, know and do.

What are you going to do when food cannot be delivered to your market or when petroleum is unavailable for periods of time or the government does not have the money to take care of the problem that is occurring in your vicinity?

These are all parts of your future world. The evidence that they are coming can be discerned without a great education. What are you going to do when you cannot move about, living out on the outskirts of the city, or up in the mountainside? What are you going to do if you cannot afford medical care and there is no one there to afford it for you?

These are the questions that are problematic and challenging. The man or woman of Knowledge considers these things because they are part of life and part of the great changes that are coming to the world.

You must not respond with panic or fear, for that will lead to bad decisions and make you ever more vulnerable. No, the response must come from Knowledge. The steps are many. If you wait till the last minute, there will be panic and chaos, and nothing can be done. So you begin to prepare. You prepare before the storm arrives.

This is life. You live in the moment, and you prepare for the future. You prepare for adulthood as a child. You prepare for family life if that is your destiny. You prepare for old age. You do not just wait until it happens. You prepare for the eventualities of life even under normal circumstances. That is intelligence. But even amongst those who are able to make these preparations, who is doing this sufficiently?

Your governments will not be able to provide endless welfare and recompense for the failure to prepare and to respond. Let your health lapse into degradation, and there will be nothing there to restore you. There will be no one there to rescue you if you fail to prepare and to respond appropriately. And your little handheld gadgets are not going to help you then. And the marvels of technology may not meet the basic and essential needs that you and everyone around you will have.

You are preparing for a world of greater difficulty, where everything will be more difficult to secure and to maintain, where resources will be limited and human ingenuity and cooperation will be called for as never before. It is like you will be living in a wartime society where everyone in your nation must pitch together and utilize every resource effectively and efficiently, except that this need will be ongoing and not simply temporary in nature.

God is giving you the warning, urging you to prepare and showing you where the source of your strength, courage and integrity resides. It will not be in your ideas. It will not be in your assumptions. It will not even be in your faith in God, for the faithful and the faithless will all perish together if they are unprepared for the Great Waves of change.

SEEING WHAT IS COMING

Unless you can receive what God is really providing you and has already provided for you, then your faith will not give you any advantage and can blind you even further, robbing you of your own responsibilities and masking your own strength and courage.

People want painkillers, but they do not choose real health, real vitality, true engagement in the world, true engagement with Knowledge within themselves. What can you do for them? What can God do for them?

Your task is to receive the Revelation; to overcome your fear and anxiety, which paralyzes you; to begin to prepare your inner and outer life; to take the Steps to Knowledge; to bring clarity and resolution into your circumstances; to depart from those who are robbing you of your inspiration.

This is what it means to respond, to be responsible. This is what God is emphasizing. That is why you must step aside from the amnesia of culture. You must be willing to break free, for humanity as a whole is heading for calamity. Those who can respond will make all the difference, not only for themselves and their situation, but for the world around them.

That is the importance of your calling. That is why you have a calling. That is why you have come into the world.

As revealed to Marshall Vian Summers, May 22, 2011

CHAPTER 11

Signs from the World

Many people today are feeling anxious about the future. They are feeling anxious about the course that humanity is taking and about the great difficulties that are emerging within and beyond nations. Perhaps their fears are associated with specific things or specific tendencies. But however they evaluate their feelings, their feelings are giving them a sign and a clue that the world is entering a more difficult and prolonged challenge and set of circumstances. Regardless of people's beliefs and assumptions, their feelings—if they can be anticipated and really considered—will give them these signs.

Perhaps people will say, "Well, I am very hopeful things will turn out well." Perhaps they will say, "Well, we will meet these challenges with new technology." Perhaps they will say, "We have been through difficult times before, and we will endure and emerge successful here."

But these are all evaluations of a primary experience that should not in any case be discounted. If people listened to their experience more than their ideas, they would have a truer sense of what is coming over the horizon. And they would have a truer sense of how they are really responding.

People do not want to suffer, and so they tell themselves things to relieve themselves of anxiety or concern. This is understandable, of course, but it often betrays a deeper experience and a real sign that one is receiving within oneself.

PREPARING FOR THE GREAT WAVES OF CHANGE

For the world is giving you signs, and the deeper Intelligence within you called Knowledge is also giving you signs. They are giving you clues, indicators and warnings. Knowledge within yourself will continue to try to flag you to get your attention. And once your attention has been gained, it can begin to guide you in a constructive manner to prepare for the changing circumstances of the world.

People rely upon their beliefs and assumptions to such a great degree that often they miss the signs, they miss the cues, and nearly everyone around them is missing the signs and the cues because they do not want to be concerned. They do not want to feel helpless or hopeless or powerless regarding the things that they are feeling and seeing. And so they try to dismiss it altogether or to lose themselves in some other activity to be constantly distracted.

This is one of the reasons that people maintain a constant level of distraction, because they do not want to be with their own experience. They do not want to feel what they feel. They do not want to feel the emptiness of their lives or the pain of compromise. They want to stay occupied at all times.

When they were babies, their parents constantly gave them distracting things—toys to play with, things to do—so that they would not cry or be uncomfortable. And now here as adults they try to give themselves distractions—things to do, things to play with—so they will not feel what they really feel.

Not all your feelings are signs, and not all your feelings, certainly, demonstrate truth. But your deeper experience floats through your feelings and emotions. Your feelings and emotions are the medium, not the only medium, but a primary medium through which Knowledge, this deeper Intelligence within you, communicates to you.

SIGNS FROM THE WORLD

People of course have lots of defense mechanisms, not only against the outside world and against threats from their environment, but they have defense mechanisms against their own experience. They do not want to feel discomfort, uncertainty or anxiety, and so they have set up a series of defenses so they can stay preoccupied and feel better about themselves and their situation.

But these defenses end up denying critically important experiences that they are having. Part of this defense mechanism is developing a whole network of fantasies that you indulge yourself in and attempt to create a happy or interesting set of dreams to replace how you are really feeling in the moment.

The truth is that many people are not feeling very good in the moment. They are not feeling very good about their circumstances. They are not feeling very good about the compromises that they have made. They are not feeling very good about their own experiences and activities. They are not feeling very good about the prospects for the future. And they are not feeling very good about what they see and feel about their future.

Here people are more perceptive than they realize. Here their feelings speak more clearly than their reactions to their feelings, or to their defense mechanisms. Here the pursuit of happiness becomes very self-deceptive because you only want to acknowledge those feelings and experiences that support your pursuit of happiness, while denying or decrying all the rest of your experience.

This leads to a kind of self-avoidance and low self-esteem because if you cannot recognize and acknowledge your own experience, then you are really denying yourself. You are in a state of avoidance with your own experience.

PREPARING FOR THE GREAT WAVES OF CHANGE

If people stayed with their experience, it would lead them eventually to seek resolution and to change their circumstances to whatever degree they can. It would make them more honest with themselves and with other people, thus enhancing their relationship with themselves and with other people.

That people have so much difficulty in being with their experience, and in learning how to communicate their experience constructively, represents a fundamental weakness in the human family—a weakness that generates immense suffering and self-disassociation, a weakness that erodes your relationship with others. For if you cannot be honest with yourself, how can you be honest with others? If you cannot respect your own experience, how can you have any respect for theirs?

Acknowledging your own discomfort here is very valuable because it is opening the door within yourself for a deeper experience to emerge. Even if your experience is unpleasant or uncomfortable, even if it is giving you a sense of anxiety about yourself and about the future, it is preparing you. This is part of what nature has provided you to survive changing circumstances and to anticipate critical catastrophes.

When great storms come in the world, the animals become quiet and seek shelter, but people pay little attention, caught up in their pursuits, in their activities or in their fantasies. In this respect, the animals are acting much more intelligently than the people.

This is a critical problem, you see, because the vast majority of mistakes that people make, they make because they are not paying attention. And the reason they are not paying attention is they are not present with their own experience, and they are not present with what is happening around them.

SIGNS FROM THE WORLD

A big part of the problem here is social conditioning. All of the training, emphasis, correction and punishment that has been placed upon you to behave in a certain way, to only show certain parts of yourself, to only express certain feelings and not others is an overarching problem that affects nearly everyone. Here your behavior is constrained, it is governed by the expectations of others, and should you violate these boundaries, you face ridicule, repudiation, rejection or in some cases severe punishment.

The only part of you here that is really free of these massive constraints is the presence of Knowledge within yourself, for it exists beyond the reach of social conditioning and beyond the reach of seduction and manipulation. That is why it represents the freest and most natural part of you.

That is why when you begin to take the Steps to Knowledge and to build a connection to this deeper Intelligence within you that God has placed within you, you begin to unwind and unravel the constraints that hold you back and that limit your experience and your expression. Here Knowledge will teach you how to express your feelings constructively in such a way that others are not harmed, in ways that lead to the greatest possibility for acceptance by others.

People do not realize how much they are governed by their social conditioning, and how through their moment-to-moment experience and behavior, they have a very, very slight awareness of what they really feel and know about things around them. This leads to a lot of compulsive behavior, social behavior, where people mimic one another to be accepted.

That is why people within their own group or set of associations talk and act and look so much alike. Here their real nature may be completely different, but they have assumed a kind of social posture and behavior for social acceptance within whatever group they identify with.

This is understandable, but it is really unfortunate because these people will not function intelligently, they will not think for themselves, and the vast majority of their real experience will be denied and avoided. They will be standing on the beach with their back to the ocean when the Great Waves arrive. They will be in the wrong place at the wrong time, unaware of what is coming. If the animals have all sought higher ground and have all withdrawn from the shore, why are the human beings standing out there? The human beings who have such greater intelligence and sensory awareness, why are they standing out there?

Why are people living twenty miles outside of the town when the world's oil supplies are diminishing? Why are people invested in things that have no future, that will not survive in the Great Waves of change that are coming to the world? Why are people not paying attention to the changing circumstances of their lives, hoping and believing everything will get better or that someone else will fix things up for them?

Clearly this is unintelligent, and yet you can see this is a product of social conditioning and social adaptation. It is the problem of human denial. It is the problem of not honoring and observing your own experience to see what it is telling you, to see the signs it is giving you. It is avoiding the signs of the world so you do not feel uncomfortable or anxious.

The world is giving you signs. It is telling you what is coming, but you must really look, and you must look even if others are not looking. You must pay attention and follow what you are looking at to learn more about it, for you cannot see the truth in a moment only.

SIGNS FROM THE WORLD

There are Great Waves of change coming to the world: environmental deterioration, changing climate and violent weather, diminishing resources, growing political and economic instability and the growing risk of competition, conflict and war [between] groups and nations over who will have access to the remaining resources of the world.

These are the great changes that are coming to your world, changes that will affect everyone, even in the wealthy nations—everyone. Why would you not pay attention to these when they have such great bearing on your life and future, and the life and future of your children and loved ones?

This requires real courage because you will have to face problems for which you do not have answers. You will have to accept that you do not really have answers, and that your ideas and your solutions are really either imaginary or insufficient to deal with the Great Waves of change that are coming.

You may deny them or dispute them or try to minimize them in your estimation, but your feelings will tell a different story. You may tell yourself anything. You may seek experts who will tell you what you want to hear to mitigate your feelings of concern, but your feelings will tell a different story.

The signs of the world and the signs from Knowledge within yourself will tell you what is coming and will indicate what you must now consider in terms of preparation.

Those who prepare for the future will be able to avoid its dangers and its dilemmas to a very great degree. Those who do not prepare will be extremely vulnerable and will have few choices when the great impacts occur in their lives. But preparation means that you must be willing to think independently, to think beyond your insecurity, to think with determination and with courage and as much objectivity as you can muster.

You cannot be panicking and make wise decisions. You cannot be governed by fear and make wise decisions. Perhaps after a period of shock, awe and dismay, you will finally settle in to a more stable and more objective perspective.

Here you will turn to Knowledge within yourself and Knowledge within others and ask: "What must I do?" Here you are willing to act because you must act. Avoidance, denial and speculation only weaken you now, only make you more vulnerable. They deny your recognition that you must prepare, and they deny that you have the strength to prepare.

God has given you a deeper Intelligence to guide you, to protect you and to give you the strength to overcome your weaknesses, to overcome your social conditioning, to overcome your need for approval and consensus with others. This is very important because you may be the only person you know who is really responding to what is happening in the world in a wise and practical way.

Here you cannot be governed by the will and the consensus of your friends and your family. You must think as an individual. You must use the power of Knowledge. You must face the world and face your own anxieties. You must get past fear and the feelings of helplessness and ask yourself: "What must I do?"

If you do not act, then your feelings of helplessness and hopelessness will increase, and you will lose the momentum that you will need to have. To be aware and not to act leads to another kind of crisis. Here people know they must get off the sinking ship, but they do not get

SIGNS FROM THE WORLD

off the sinking ship. Here people become frozen, unable to act, in the face of immediate and tremendous danger.

If you are aware of change and the need to prepare, but you delay this preparation, then your choices become more limited and more costly. You do not want to wait [until] everyone becomes aware because then it will be panic and very little will be available to help you.

Before you can help others, you must secure your position. Before you can help those in need, you must become strong and secure in your position. You cannot achieve ultimate security because that is not possible in the world, but you must have a stronger position. If you live twenty miles outside of town and there are shortages of fuel, you will be stranded. If those conditions continue, you will be a refugee.

It is perhaps unthinkable in the wealthy nations that this could take place, but that is because people think the future will be like the past. That is because of their social conditioning and their beliefs and assumptions, but that is not what their deeper feelings are telling them.

Look over the horizon of your life and see what is coming over the horizon and ask yourself: "What do I really see coming over the horizon?" This is a different question than "What do I want to see? What do I hope to see? What do I think I want to see? What do I believe will happen?" It is not about hope or assumption and belief. It is about seeing.

When the Great Waves come, why are the people standing on the beach? Why are they acting like nothing is happening when all of the animals seek higher ground? The stupid animals, why are they more intelligent than the people?

PREPARING FOR THE GREAT WAVES OF CHANGE

What is coming over the horizon? What do you really see and feel? Forget what you want to see. That is foolish and self-deceptive—what you want. What do you really see? And what must you do to respond to this?

Look at your situation: where you live, how you live, how you travel about, the nature of your work, your circumstances and your obligations to others. It will all have to be re-evaluated in the face of the Great Waves of change. If you do this in advance, then you can make the necessary adjustments and create a much stronger and much more resilient position for yourself and your family.

If you become aware of great change and do nothing, your anxiety will increase and so will your sense of powerlessness. You will be held in place, hoping that someone will take care of these problems, hoping for a miracle from God because if awareness does not lead to action, it leads to a sense of powerlessness and depression. If the person standing on the beach waiting for the Great Waves to come thinks to themselves, "Maybe I should get to higher ground?" and they do not move, they are lost.

Knowledge within you is telling you where to go, what to do, what to give yourself to, what not to give yourself to, where to move forward and where to hold back. It is doing this on a daily basis. But most people are far too preoccupied and caught up with themselves and have too much resistance to their own experience to hear and follow this natural inner guidance that the Creator of all life has given them.

This is the great endowment within you. But it has no value if you are defending yourself against it, if you are distracted continuously, and if you are preoccupied with other things.

SIGNS FROM THE WORLD

Nature has equipped you to deal with uncertainty, to deal with catastrophe, to deal with hazardous events, but so many people have lost this natural capacity. They build their lives in the most vulnerable places, under the most vulnerable circumstances. Many people are forced to do this, of course, because of poverty and political oppression, but that is a different matter. The tragedy here is in the wealthy nations where people have choices, where people can determine what they will do to a far greater degree.

Look at how people live and where they live. They are heedless of the powers of nature. They are not paying attention to the Great Waves of change that are coming to the world. They are not looking, listening and paying attention. And if they see potential problems, they think it is not a problem for them, that society and civilization will solve all the problems, so it is not a problem for them.

Here sometimes people see things that they need to see, but they do not respond appropriately. They respond foolishly. They will build their house on the beach. They will live twenty miles outside of town. They will build a career that has no future and no foundation. And when the difficulties arrive, which they will, they will say to themselves, "I did not see it coming."

They did not see it coming. It was coming, and the signs were there, but they did not see the signs. They would not heed the signs. They did not head for higher ground when they had time to head for higher ground, and now they are standing on the beach and the Great Waves are coming.

They will go down with the ship because they did not respond to the signs. And they could not overcome their own feelings and experience of helplessness and confusion. They will be overtaken by fear, they will be immobilized and unless someone shakes them out of their dream, out of their frozen state, then they will just go down with the ship.

You should not think that this was their destiny or that God was calling them back or that it was their time to leave the world. Those are all foolish excuses to cover up the mistakes that people make.

Knowledge within you wants to keep you alive so that you have a chance to fulfill your destiny in the world. It will do everything it can to gain your attention and to guide you. If you step in front of the truck, it is not because Knowledge took you there. If you cross the street speaking on a telephone and are struck by a car, it is because you are not paying attention.

You are not using caution, because the world is a hazardous place, and you must use caution. When you step outside the door of your house or the building you are in, you must use caution. Be observant. Be wary, just like all the creatures of nature are wary to the extent of their awareness. This is not being fearful; it is simply being objective and observant.

If your economy is going to fail, you will see the signs. If you are paying attention, you will see them. If you respond appropriately, you will prepare yourself for these events.

God is warning you. The world is warning you. It is telling you what is coming, not specifically, not by a date. It has nothing to do with dates. It is the movement of life. It is the changing circumstances of your life. Ignore them at your peril. Deny them at your peril.

You must always be watchful. That is part of being aware; that is part of being alive. This world is not your heavenly state. You must be

SIGNS FROM THE WORLD

careful here—not fearful, not living in trepidation, but watchful and observant.

If people would do this, it would eliminate the majority of their errors. And if they responded to what they were seeing appropriately and stayed with their experience to consider what it really means and what it might require of them, they would be in a much better position than they are today. If they referred to Knowledge in terms of their decisions about who to be with and what to do, the world would look different and feel different than it does today.

Listen to your experience. If you are feeling constant anxiety or reoccurring anxiety, or reoccurring insights or reoccurring ideas, then pay attention to this. There are three stages here. There is the awareness of something. There is being with that awareness to study it and to try to understand what it means. And then there is taking action. There is seeing the sign, there is being with the sign to see what it means and what it has to teach you, and then there is taking concerted action.

Action here might require many steps; it is not just one action. If you realize that where you live is not sustainable into the future, that will take a whole series of actions over time, and you will carry out these actions in stages. If life is telling you to move away from future danger, that might require many actions, a whole plan of actions.

Life will give you this time, but you must respond. You cannot deny or delay what must be done. Advancement here means that the time between seeing, knowing and acting is diminished. The time it takes between seeing that your marriage has lost all of its life and purpose and knowing what this means and taking action, shrinks. The time it takes between seeing that you must exercise and take care of your body, knowing what this means and taking action, shrinks.

Then when faced with immediate hazards, you can see, know and act appropriately. You will not be frozen in confusion. You will not be thrown into suspended animation. You will not be locked and tied, held down by your social conditioning or by your fears or sense of insecurity.

God has given to you the power to do this. It is the power of Knowledge. That is why learning The Way of Knowledge and taking the Steps to Knowledge build this core strength and set of abilities. This restores your relationship with yourself, your ability to function successfully in the world and your ability to be with the experience of others—to learn from their successes and their failures, to honor their humanity and to support the emergence of Knowledge within them.

While this might seem to be common sense, it is really revolutionary because it shifts your relationship with your mind, with your body, with your culture, with your family and with the entire world around you. It is a great shift of authority from your ideas and social conditioning to a deeper authority that God has placed within you, an authority that cannot be fooled, an authority that has immense power and capabilities.

Humanity is entering a time of unprecedented and escalating change. How will it respond? Will it prepare wisely and in a timely manner? Will it choose to cooperate amongst its nations and peoples? Or will it fight and struggle over who will have the last of the world's resources?

Humanity is growing in size, but it is drinking from a slowly shrinking well in the world. How will it prepare for this? Can it become aware of this? Will it understand what this means? And will it act appropriately and constructively? These questions begin with you and how you respond to your own life and circumstances.

SIGNS FROM THE WORLD

You make your contribution first by bringing wisdom and competence into your own life. Then you will be able to assist others because you yourself will have seen the power of Knowledge, and you will see the obstacles that stood in your way and how they could be overcome.

Instead of dwelling over what will save humanity, you had better concern yourself with what will save you and those that you love and who you are responsible for. Knowledge will give you the awareness and the pathway to follow. And it will break the chains of social conditioning and the chains of weakness and avoidance within yourself, giving you great strength and self-assurance.

You must bring strength to the human family, not weakness. Your part to play is to find this strength within yourself, to apply it and follow it in your life and to contribute it to others. This you do by demonstration more than by precept.

It is this that will show others that they too have the power of Knowledge—they too have the power to overcome the inhibitions that exist within themselves to face and to prepare for a future that will be unlike the past.

As revealed to Marshall Vian Summers, November 7, 2008

CHAPTER 12

Self-Reliance

Humanity is entering a time of great difficulty, a time where you will face the Great Waves of change that are coming to the world: environmental decline; the depletion of resources; growing political and economic instability; and the increasing risk of competition, conflict and war between nations over who will have access to the remaining resources of the world.

It is a time of increasing instability and uncertainty. Governments will appear to be unable to deal with the situation as it unfolds. People will be restless and anguished and angry at the corruption and mismanagement that will emerge.

It will be a time of increasing friction between nations that do not have good relations with one another. It will be a time when poverty and deprivation will grow because the production and distribution of food will be disrupted.

It is a time for humanity to become strong and united in order to weather the Great Waves of change and to forge a stronger union between peoples and nations for the preservation of the world and of human civilization.

Most people really do not see the magnitude of what is coming. Perhaps they feel a deep anxiety. Perhaps they feel a sense of trepidation. Perhaps they are concerned with their own economies and their own national problems. But the specter is far greater. The challenge will be far greater.

People's confidence in nations and in leaders will be shaken. People's confidence in God and in God's Providence will be shaken. Some people will think that this is the end of time, the great end of time that has been so anticipated by certain religious groups of people.

But it is really a time when humanity must face the consequences of its misuse of its natural inheritance in the world and its greedy and exploitive use of the world. It is a time when the consequences must be faced, when the bill has come due, when humanity must reckon with the results of its behavior over a long time.

This, however, is a very important time for you. For it is a time of re-evaluation of your priorities and your activities, your relationships and your commitments. It is the time to build a greater self-reliance, not a reliance on your ideas or beliefs, not a reliance upon your assumptions, but upon a deeper power that the Creator of all life has placed within you, a reliance upon the power of Knowledge—a deeper Intelligence, an Intelligence born of God, an Intelligence that is not a product of your social conditioning or preferential thinking.

You see, you really have two minds. You have a surface mind that has been conditioned by the world and is a product of your adaptation to the world and your social conditioning. And then you have a deeper Mind of Knowledge.

Knowledge does not think like your personal mind. It does not compare and contrast. It does not speculate. It is not conniving. It does not bargain. It does not involve itself in endless conversation with other people. It is entirely different, you see.

SELF-RELIANCE

So when We speak of self-reliance, We speak of your reliance upon the power of Knowledge. For it is this power that the Creator of all life has given to you to guide you, to protect you and to prepare you to live a greater life in a changing world.

Unlike your personal mind, Knowledge is not afraid of the world. It has come here on a mission to serve the world in a unique way, involving certain people and certain situations. It is not paralyzed by fear. It is not dominated by denial or preferential thinking. It is clear, honest and objective, and it sees through deception of every kind.

You cannot use this greater Intelligence to enrich yourself as if it were merely a resource for your personality. You cannot use it to gain wealth and advantage. For Knowledge is much more powerful than your intellect and functions beyond the realm and the reach of the intellect.

It is very important here that you realize the distinction between Knowledge and the mind where you live 99% of the time. Otherwise, you will misconstrue this message. You will begin to rely upon your assumptions, your beliefs, your attitudes and your prejudices. This will only weaken you and blind you and make you less able to adapt to the changing circumstances of your life. It would make you unable to see what is coming over the horizon: the Great Waves of change. And it would not enable you to prepare for the Great Waves of change.

Therefore, the distinction between Knowledge and your personal mind is essential. But this distinction must be based in experience. For when you are connected to Knowledge, it is a very different experience than simply making decisions in your mind, or trying to adapt your thinking alone to the circumstances around you.

PREPARING FOR THE GREAT WAVES OF CHANGE

For when you try to adapt your thinking, you are always referenced to the past because your mind is always referenced to the past. While history may teach you, and does teach you important lessons, to be so past referenced does not enable you to see a situation in a new way. It does not enable you to discern a decision you have to make or a phenomenon that is occurring around you, for you will always think of it in past terms.

This is like driving the car looking through the rear window. You cannot see where you are going, you cannot evaluate changing circumstances, and you cannot recognize people and their abilities and disabilities effectively because your mind is so conditioned to think along certain patterns that it cannot be creative or discerning in its outlook. That is why your intellect must follow Knowledge, for Knowledge will never follow your intellect.

Another way to think of this is to say part of you is Divine and part of you is human. Part of you is wise; part of you is not wise. Part of you cannot commit errors; another part of you is very prone to committing errors.

So which part of your mind should follow the other? And how do you cultivate your awareness of Knowledge, develop a foundation of experience in Knowledge so that you can realize how it functions through your unique nature, and how you can read its signs and its direction?

Self-reliance, then, is not something that just happens. It is not something that happens by accident. It is something you must cultivate and develop. At the very center of your life, this must be your focus. For if God has given you the guiding power of Knowledge to enable you to discern your environment and to make wise

SELF-RELIANCE

decisions, how could this not be the center of your attention and your primary focus in education?

The universities may teach you about accumulated understanding and wisdom from the past, but they cannot teach you how to see and to know and to act with wisdom.

Old ideas will not fare well in the face of a radically changing world. Old assumptions will prove to be either completely false or highly inadequate to meet the situation at hand. Do not look at the changing circumstances of your life and think it is merely a cycle like the past, an economic cycle or a life cycle or a sociological pattern alone, for the world is changing and has changed.

Humanity has altered the climate of the world. Humanity has despoiled the waters of the world. Humanity has ruined the soils of the world. You will now have to enter a declining world, a world where the resources will become more difficult to acquire and far more valuable.

This will change the behavior of peoples, especially in the wealthy nations. You will be adapting to a whole new set of circumstances where your past training and understanding may prove itself to be not very useful.

It is those who can see the Great Waves of change, who can recognize them in advance and have the courage to prepare for them, and have the courage to teach others how to see and how to respond to the world that you have come to serve.

Knowledge is here on a mission. It is here to serve. Its focus is not simply acquisition, security and enrichment. It is not impressed by beauty, wealth and power. It sees how humanity enslaves itself and deceives itself and degrades itself. And it sees the terrible consequences of human poverty and deprivation.

Knowledge is here to renew your life, to restore your integrity and to re-establish the Divine Presence and Power in your life. This happens no matter what religious tradition you might be part of. It happens even if you do not have a religious tradition that you can define.

It is the power of redemption within you. It is not governed by your intellect or by your social conditioning or by the opinions or the values of others. It is not governed by your religious ideology or beliefs. In fact, these things often prove to be impediments to the recognition and to the expression of Knowledge.

God has given you great power, but it is not your personal power. It is not the power of your will or self-determination or domination of others. But to know this and to feel this and to see the evidence of this everywhere around you, you must engage in a different kind of education, an orientation within yourself.

That is why the Creator of all life has sent the Steps to Knowledge into the world so that you can make this Divine connection and learn to realize, to recognize, to experience and to accept the power and the presence of Knowledge within yourself, and this power within others.

This will alter your perception and your understanding because Knowledge will give you a real understanding. It will temper your ideas, and you will see that your beliefs and your ideas and your assumptions are all relative and are all referenced to the past, and how much of your thinking is based upon fear and preferential thinking, which in most cases is merely a response to fear. You will need now to learn to listen within yourself as well as looking about with great objectivity.

SELF-RELIANCE

Most people are dominated by their minds, their thinking, their planning, their fears, their worries, their issues, their anger, their unforgiveness. They are present in the world, but they are not present to the world. They are standing in front of you, but they are consumed with their own thinking and cannot see and cannot hear and cannot respond appropriately to their environment.

As you watch people, you will see how consumed they are in their own minds—trying to have what they want, trying to escape loss, trying to impress other people, being driven by desires, fantasies, fears and needs. To the person guided by Knowledge, it will seem like everyone else is in chains even though they may be trying to live in splendor, or live in a politically free nation.

Knowledge will give you the eyes to see and the ears to hear. It will give you the strength to face anything and to navigate what seems to be utterly unpredictable and overwhelming.

No matter what your beliefs or ideology or social background or national heritage, Knowledge is alive within you. It is trying to steer your life in a positive direction and to keep you from making self-destructive decisions. You can feel its urgings and its restraints if you are really paying attention.

But this takes a very different approach. Here you must become more of an observer of your own mind and the world around you rather than one who simply reacts to the world and condemns the world as a result. Here you must be willing to live with questions and not be relying upon answers. Here your mind must be open like a window, instead of closed like a door. Here you must have the courage to re-evaluate your obligations, your relationships, your activities and your commitments to see if they really are part of your greater purpose in being in the world.

PREPARING FOR THE GREAT WAVES OF CHANGE

To your intellect, this may be very confusing, but to Knowledge it is clear as day. As you begin over time, by taking the Steps to Knowledge, to see the difference between how Knowledge functions and how your intellect functions, you will see that they are entirely different. And you will come to realize that you could never figure out in your mind the real truth of your life, the real direction of your life and how to be in the world successfully, where your greater purpose can be realized and fulfilled.

Look around and you will see that people are still trying to acquire ever more. They do not realize they are living in a world of shrinking resources. They are unprepared for what is coming. And when the Great Waves hit, when economic hardship occurs suddenly, it is as if they did not see it coming. Even the experts did not see it coming.

But Knowledge is giving you signs and clues. The world is giving you signs and clues. But if your mind is not paying attention, if you are not watching and listening, without constantly judging and interpreting, then you will not see and you will not hear. And the signs will be lost upon you.

Your self-reliance, then, must be reliance upon Knowledge. And you must have a very clear understanding here that who you are is not your mind, that your intellect really is a vehicle of communication in the world to serve a greater power within yourself and to serve the Greater Power that has sent you into the world.

With Knowledge, you will know what to follow within yourself, and you will know who to follow around you—who is strong with Knowledge and who is not, who can see clearly and who cannot, who has humility and who does not, who is driven by ambition or social conditioning and who is not. This gives you a basis of discernment in relationships, which is one of the critical skills you really must develop in life to be successful and to not give your life away to people or to circumstances that have no future or real destiny for you.

You will find that Knowledge does not look upon the world with horror or with condemnation, or with frustration or anger. It does not look upon the world in a fanciful way either, hoping or believing that everything will be fine.

It does what your intellect cannot do, which is to see clearly. Knowledge is here on a mission. It is not here to condemn the world, it is not here to enrich itself, it is not here to manipulate other people, and it is not here to gain power and dominance. You must see that this power cannot be found anywhere else.

Then when people are guided by Knowledge, their thinking begins to represent Knowledge. Their values and their attitude change. They are not driven by greed. They are not mesmerized by beauty or wealth. They do not look upon life as simply an opportunity to acquire wealth or power for themselves. There is something rare and unique about them. There is a presence with them. You feel confidence in them. You trust their wisdom, and you recognize their humility.

Knowledge is the greatest power in the universe because it is the Power of God working through the individual. Yet people do not know of Knowledge. They do not respond to Knowledge within themselves. They do not follow Knowledge in their decisions and behavior. It is as if Knowledge either did not exist, or was some very mysterious thing that only happens to certain people.

But this is not the case, you see, for everyone and you were sent into the world for a greater and unique purpose. But only Knowledge knows this purpose and holds this purpose for you—waiting until you are ready, waiting until you have the maturity and the desire and the commitment to discover and to experience your greater purpose in being here.

Sometimes Knowledge will have to wait a lifetime, for people can continue doggedly in their old ways of thinking seemingly forever—unwilling to question themselves, unwilling to reconsider their opinions and their judgments, unwilling to face uncertainty, unwilling to yield to a greater power.

When Jesus said, "Come follow me," he meant, "Follow the power of Knowledge, for that is what I am following." Here God is giving you great power, but also great responsibility. You may pray to God to fix the world, but God has sent you here to fix the world, you and everyone else.

You may have personal reasons for being in the world, but there are greater reasons for you being in the world. And your greater purpose does not mean you will become an ascetic or a renunciant, for that is only reserved for certain individuals. For everyone else, they must be in the world, functioning in the world, dealing with the world, serving the world.

Who will restore the environment? Who will change their lifestyle to be able to function in a declining world? Who will preserve the soils of the world or the waters of the world so that humanity may have a future here? Who will meet the great human need and face a world where poverty will grow? Who will have the strength to face the Great Waves of change and to use them to be of benefit to the world? You cannot look to governments or leaders to provide all this. If you do, you will be gravely disappointed.

SELF-RELIANCE

You see, the answer is within you, but the calling is in the world. The world must call out of you your greater gifts, which means you must face the world and not condemn it. You must be open to the world and not be retreating from it, or running away from it.

So very different this is, you see. It is a different incentive. It is a different way of being in the world. It is a different relationship with yourself, most fundamentally, and it establishes a very different foundation for relationships with others.

For with Knowledge, you will know who to be with and how to be with them. And Knowledge will take you to the individuals who will have the greatest impact and benefit for your life. Knowledge is not driven by insecurity or the need for wealth, beauty or power, so it will not be deceived by these things, and it will free you from being deceived by these things.

Here self-reliance does not mean you go it alone and you do not trust anyone else. No, indeed, Knowledge is here to unite you with certain people, to unite you with your purpose for being in the world, to bring that love and presence and service to the world. It is to re-engage you with life, to unite you with others, to liberate you from Separation and from the hell of your own isolation.

Only Knowledge has the power to do this, you see. Your mind may conjure up wonderful and fantastic belief systems, or metaphysical theories, or religious ideology, but it cannot do what Knowledge can do because your mind cannot do what only God can do. And God reclaims the separated through Knowledge.

It is not just the Knowledge of God, for you really cannot know God from your current position. It is by following what Knowledge has come into the world to accomplish that reunites you with your

PREPARING FOR THE GREAT WAVES OF CHANGE

Source, that liberates you from the tragedies and seductions of the world, and that gives you a real relationship with yourself and with others, relationships that cannot be established in any other way.

This goes far beyond intuition, fleeting moments of intuition. For Knowledge is a power within you that will emerge in your life as you create an opening for it, and will give you an entirely different experience of yourself and of the world around you.

God has already answered your prayers, even if you have not yet prayed. God has answered the need of the world. It lives within each person, waiting to be discovered.

At the outset, you will have to break free from your social conditioning, from your habitual patterns of thoughts, from your sense of obligation, from your sense of powerlessness. You may have to change the circumstances of your life, alter your commitments to other people, break free of family obligations to a great degree, break free of your own fixed ideas and your identification with those ideas.

For God's first purpose is to unburden you, to free you, to teach you what real freedom is, to show you what is holding you back, what is holding you in chains and enslaving you. You cannot serve a greater purpose or a greater power if you are bound to other things.

You must trust what is mysterious and deep within you that exists beyond the realm and the reach of your intellect. And you will have to break the chains of obligation, the chains of submission, the chains of trying to please others, or to uphold the values of your culture, if they are not in keeping with Knowledge.

This is the real form of self-liberation. This is the power of redemption working within you—beyond your control, beyond

SELF-RELIANCE

your understanding, beyond the boundaries and the restraints that culture and even religion have placed upon you.

Here you break free so that you may unite with the Power of God within you and to bring that power and that presence into your life and into the world. You do this without espousing religious beliefs or promoting religious institutions. It is the presence that is with you that is the most important thing. And if you are not dominated by your thoughts, your fears, or your desires, you will feel this presence, for the great well of silence exists beneath the surface of your mind. It is there that this presence dwells.

Break through the surface and you will feel it. But if you live at the surface, you will never know what is beneath you. You may float a little boat on the ocean and never know what is beneath you. What is really valuable is beneath the surface of the mind: the strength, the power, the grace. That is where it is. You cannot get there by trying to just use willpower, or think along certain lines, or adopt a certain ideology.

You must learn to be still. You must learn to refocus your mind, to gain authority over your emotions, to bring the wild horses of your thinking to serve you and to serve Knowledge.

There will be many questions of course. There will be much confusion. There will be much re-evaluation.

But as you take the Steps to Knowledge, you will see how enslaved you were before, how weak was your position, how you could not see or recognize the signs of Knowledge within yourself or within other people, except perhaps on very rare occasions. You will see how much you were a slave to your needs, your fears and the expectations of others.

PREPARING FOR THE GREAT WAVES OF CHANGE

But the process of liberation must occur. But this liberation is not just to set you free. It is to put you in a greater role and set you in a greater direction that represents your purpose and destiny here.

The world is chaotic. It is not governed by God. But God has given you a power to navigate it, a direction within it and a greater purpose that is for you to serve. That is why you have a unique nature and personality because you have a design for this purpose.

Come to recognize your design and you will come to experience your Designer. You are uniquely made for a purpose that you have not yet discovered. So how can you judge yourself?

You will need Knowledge now as never before, for you cannot rely upon anything around you to be safe and secure in the face of the Great Waves of change. You cannot fall back into laziness or indolent thinking and find any stability or security there.

Here the Great Waves of change will serve you in requiring you to develop this self-reliance, this reliance upon Knowledge. Here what seems terrible and traumatic, dangerous and overwhelming, will call Knowledge to come forth.

You cannot be fooling around in the face of the Great Waves of change. You must begin to take your life and your time seriously and not squander them on meaningless things. In this, the Great Waves are a calling—calling for you to respond to Knowledge within yourself and to the power of events in your world. Here you will see that the world is serving you, trying to prepare you, trying to awaken you, trying to call you out of your dreams of fulfillment and tragedy into a greater service and participation in life.

SELF-RELIANCE

Here you must forego your condemnation of the world, your repudiation of the world, your escape from the world and find the strength to face the world with clarity, objectivity and humility. Knowledge will give you the power to do this, for Knowledge alone has the power to do this.

The stronger Knowledge becomes within you, the more it will resonate with other people, and the more you will become a source of strength and inspiration, and your life will be a demonstration of a greater service in the world. And that is the most powerful spiritual teaching there can be.

Here you will be following the same Power that guided the Jesus and the Buddha and the Muhammad and all the saints that have come and remain unrecognized to this day.

Your mind will take issue, your mind will question, your mind will doubt, but that is only your social, surface mind. Of course it will respond in these ways, for it is profoundly insecure and unsure. Its beliefs are only a crutch. Its fundamentalism is only a replacement for Knowledge.

For only Knowledge knows. The mind believes, the mind tries to define and to determine life, but only Knowledge knows. That is why it is the source of your redemption. That is why it is the Power of God in your life. And that is why it is the greatest Power in the universe.

As revealed to Marshall Vian Summers, January 21, 2009

CHAPTER 13

BUILDING YOUR ARK

Great Waves of change are coming to the world, Great Waves of change that will change the landscape of the world for everyone—great change that is the result of humanity's misuse and overuse of the world, great change that will bring about a complexity of effects that no one can discern fully in the moment.

The Great Waves of change are already beginning to affect the world's economies, leading to ever-greater instability and uncertainty. The resources of the world are becoming depleted, and humanity has not prepared for its future, its future living in a world of declining resources and of environmental limitations.

Many people are filled with hubris that technology or ingenuity will overcome all of these problems. They think that they can control nature, using technology and ingenuity, and it is only a matter of investment in these things that will determine the outcome. But technology has its limits, and in the end the power of nature and the limits of the world will be the defining elements for humanity, for the future of the human family.

If you search your heart, rather than appealing to your ideas, you will feel that great change is coming. Perhaps you cannot describe it or define it. Perhaps it is unclear to you. But what is important is you are feeling the movement of change, the impending change in the world.

PREPARING FOR THE GREAT WAVES OF CHANGE

This is very important, for now you must pay attention to your environment ever more closely. You must pay attention to changing climate and violent weather. You must pay attention to the growing economic and political instability around the world. You must pay attention to problems in developing and distributing food and water. And you must pay attention to the essential energy resources upon which humanity now depends and will depend for the foreseeable future.

This is not a time for false self-assurances. This is not a time for ambivalence or complacency. It is a time to be aware and observant. Take all of the energy you spend in condemning others, in criticizing governments, in repudiating leaders and bring this energy, this time and focus to bear upon the changing circumstances of the world. Do not be concerned that you do not have an answer, for no one has an answer. There are many answers to specific aspects of the Great Waves of change, but no one has an answer for the entire thing.

The Great Waves of change are all converging at the same time, creating crosscurrents and a complexity of events that no one can predict accurately. Climate will affect food production. Food production will affect political and economic instability. Political instability will affect energy production and distribution. Poor living conditions will foster disease and even pandemics. Water resources will affect vast populations, forcing mass migrations, creating ever-greater political and economic tensions and instability. If this goes unchecked, then you are facing a different kind of human conflict. You are facing wars of desperation, which are unlike the conflicts that humanity has engaged in in the past.

The dangerous outcome can be terrifying and seemingly overwhelming. But your task as an individual is not to try to come up with an answer for everything. Your task is not to fall into despair

BUILDING YOUR ARK

and complacency. Your task is to discover the power and the presence of Knowledge within you.

For God has given you this power and this presence to guide you, to protect you and to lead you to a greater accomplishment within the changing circumstances of the world. For, you see, the Great Waves of change are not merely a big inconvenience for you. They have everything to do with why you have come into the world. The real gifts that you have to give to humanity, gifts which you have not yet discovered, will be stimulated and called forth from you by these Great Waves of change.

So denying the Great Waves of change or disputing their reality or avoiding them is destructive to you for many reasons. First, it does not allow you to prepare for the Great Waves of change. But even beyond this, it denies the possibility that your greater calling will emerge in life and with it the fulfillment that this calling will produce for you and for others.

Perhaps your first response to the Great Waves of change will be fear and trepidation, but this must be replaced by a kind of determined objectivity. For there is nowhere to run and hide from the Great Waves of change, and denial or avoidance only weakens your position, making you more vulnerable and less able to deal with the impacts of these Great Waves of change as they occur.

So you really have no alternative but to stand and to prepare. Once you realize this, your fear and trepidation and feelings of helplessness will be replaced by a kind of determined objectivity. The deeper Intelligence within you, called Knowledge, will give you this strength—for it is not afraid of the world, and it is not afraid of the future. Its concern and responsibility is to bring your awareness to a deeper place within yourself, where the power and presence of

PREPARING FOR THE GREAT WAVES OF CHANGE

Knowledge can guide you, protect you and enable you to navigate the uncertain and difficult times ahead.

So here fear is replaced by a kind of clarity and objectivity. If you are terrified, you will blame others, and you will spend all your energy blaming others, when in fact you need this energy, this time and this focus to prepare yourself and to assist others in preparing as well.

This Message is a gift of Love, to alert you, to prepare you, to warn you and to preserve you. Do not consider it as a fearful message. It is like your neighbor coming to tell you that your house is on fire. Would you accuse them of giving you a fearful message or being based in fear?

A great warning is being sounded around the world. It is a gift from the Creator of all life, warning people about the Great Waves of change that are coming to the world, and about humanity's encounter with intelligent life in the universe, humanity's encounter with a Greater Community of intelligent life. These two great events—the Great Waves of change and humanity's encounter with the Greater Community—will determine the fate and the future of every person living in the world.

There is no problem, there is no challenge, there is no priority that is greater than your focus on these two great phenomena, for they are the greatest events in human history. To avoid and to deny them is to avoid and deny your place in that history and your contribution to that history.

So when you realize there is no place to run and hide, either physically or psychologically, then you can stand firm. And you ask of Knowledge within yourself: "What must I do?" And Knowledge will begin to give you steps to take. Instead of answers or solutions,

BUILDING YOUR ARK

it gives you steps to take. You begin with what you have to do right now. You begin to resolve the problems that are robbing you of focus, energy and time right now. Perhaps there are a hundred steps that you must take, but you can only take them one at a time.

People want solutions. They want answers. They want reassurances because they are too afraid to realize that they must prepare and that the responsibility is upon them as well as upon leaders of nations and institutions.

If you do not claim your own responsibility here, you will find that you have no power in the matter, and you will feel impotent and helpless in the face of the Great Waves of change. Here you must build your own ark. And like the great story of the ark, you must build this before the great change comes. For it was not raining when Noah built the ark, and things will not be collapsing around you when you build your ark. But build it you must.

Here you do not build an ark simply to provide safety and security for yourself, but more significantly to put yourself in a position to be of service to others and to allow the Great Waves of change to call from you your greater service to the world, and to provide you the power and the inspiration to provide this service where it will be most needed and effective.

Your ark is small. You cannot stockpile food for the rest of your life. You cannot stockpile anything for the rest of your life, so that is not the emphasis. You will need a certain amount of resources available to you so you do not fall prey to the shocks that the Great Waves of change will bring. But you cannot provide safety and security for yourself forever, so the real ark building is more internal than it is external.

PREPARING FOR THE GREAT WAVES OF CHANGE

Yes, you should have three months' worth of food. And you should have three months' worth of financial resources at the minimum. Yes, your health must be checked and it must be strong. Yes, you must be with people who can assist you in supporting you in building your ark, and in building their ark as well. Yes, you must leave divisive relationships that have no future and no promise. Yes, you must re-evaluate your life, your involvements, your possessions—everything.

This is building the ark. This is determining what is essential and what is not essential, what you will really need and what you will not really need, what you can actually do and what you cannot actually do. This is why the preparation is more internal than external.

People will do many reckless and foolish things in the face of the Great Waves of change. Some people will try to move out into the country, thinking that they can grow their own food and be independent in this way. They do not realize the danger of isolation here. Some people will try to stockpile food and resources for a long time, not realizing that if they do this, that others will simply come and take these things away from them. Some people will think they should go out and build a spiritual community somewhere and to try to achieve self-sufficiency in this way, but they do not realize that if they were to do this, they would have to deny others food and shelter. For how will a spiritual community exist around such great human need and suffering, and how will you keep people from entering your sanctuary?

There are no easy answers. There is only the intention and the will to prepare. The preparation has many steps, which you cannot discern at the outset. You have to follow the greater power of Knowledge within yourself, and the power of Knowledge within others.

BUILDING YOUR ARK

In the future, there will be many dangerous and foolish leaders. They will lead humanity into ever-greater danger and despair. In the future, there will be leaders who will arise who could become vicious dictators as the human need escalates, as resources diminish, and as social disorder increases. Be very careful, then.

Here you must be able to discern the presence of Knowledge and distinguish it from the assertions of people's beliefs and ideologies. Here you must not think that all the solutions are there waiting to be employed, for this is not the case. Here you and others must work diligently for a better future. And you build your ark to give you enough stability to weather the shocks of the Great Waves of change so that you can be in a position to be a contributor and not merely a victim or a consequence of the Great Waves.

You need to have a strong mind, a strong will, a strong body and great compassion. You will have to know who to serve and who not to serve, where your gifts must be given, where they cannot be given, when to say yes, when to say no. You must break free of your social conditioning that makes you think like everyone else, because everyone else will not prepare, and if you do not prepare, you will share their fate.

This is a calling for you. Do not look to everyone else. You must look to yourself first. God calls to you to prepare, to build your ark, so that you may have a future, so that you may serve others in need and to support others in building their ark so that they may be stable and secure and strong in the uncertain times to come. Here you must listen to the power and the presence of Knowledge over the rancor and the anger and the turbulence around you and other people.

Support good leadership for your nation, but do not think that one person alone or one group of people are going to offset the impacts of the Great Waves of change. Do not take the false comfort in thinking there are just a few solutions to the whole problem, for it will take a thousand solutions to prepare for, to mitigate and to adapt to the Great Waves of change.

Here you must find your strength, your courage and your determination. You cannot be weak and helpless, waiting around for someone else to take care of these problems for you. Knowledge will give you the strength, the courage and the determination. And that is why taking the Steps to Knowledge is so essential.

For you must ask yourself: "How will I find the strength to face the Great Waves of change? How will I know what to do within the changing circumstances of the future? How will I know which way to go when everyone else is panicking, or when everyone else is responding in certain predictable ways? How will I know who to trust and what to trust within myself? How will I know what to do when no one knows what to do?"

These are all very essential questions, but they are not the questions that your intellect alone can respond to and to answer effectively. For they require a deeper power and a deeper presence within you—a presence that is not governed by fear, that is not conditioned by social pressures and expectations; a power that is not dominated by the will and the intentions of others.

You cannot find this power in the intellect, for the intellect is largely the product of your social and biological conditioning. The intellect here can only serve and be a greater resource for Knowledge, for Knowledge cannot be a resource for the intellect.

When you understand this, then you will be able to see more clearly and more effectively where your power resides, and the degree of

BUILDING YOUR ARK

responsibility you must assume to have your life have real purpose, meaning and direction. This is not simply a good idea now. It will be essential for your survival and your well-being.

For you will have to face change at a level that you have never had to face before. You will have to take risks. You will have to make important decisions. You will have to discern other people very carefully. Your mind will have to be clear. You cannot be overtaken by your own internal conflicts. Where will you find this strength, this power and this ability? Here you must turn to Knowledge, for this is the power that God has given you. And Knowledge will know what to do no matter what is going on around you.

But you cannot sit around waiting for Knowledge to guide you. You must begin your preparation. Do not wait until the skies are darkened, until events are happening so fast that you cannot respond, for then you will have no options, and you will have missed your window of opportunity to prepare for the Great Waves of change. Do not think you have all the time in the world, that you will do this next summer or next year or some later time, for you do not realize how precious this time is right now—that this is your window of opportunity, and it will not last forever.

Here you must be willing to act when others are not acting. You must decide when others are not deciding. You must let things go when others are not letting things go. You must rearrange your life when others are not really rearranging their lives. You must prepare, even though others are not preparing. You must find people who can help you do this, and you may have to leave behind or depart from people who cannot or will not help you do this.

Do not think for a moment that everything will turn out fine, that governments or technology will take care of all these problems. Do not give yourself this false assurance. You must overcome your own laziness. Your must overcome your own weaknesses here. You must not listen to those voices in your mind that tell you that you cannot do anything and that other people will take care of things for you, like you are a little child, for this will not be the case. Governments will not be able to take care of everyone in the future. Governments will be struggling just to provide the basics and to keep human civilization from falling apart.

You will not find this strength in your emotions, or your ideas, or your beliefs, for they are not strong enough or resilient enough to face the Great Waves of change. You must turn to the power and the presence that God has placed within you, and learn how to still your mind so that you can feel the power and presence of Knowledge deep within you. Over time, you will learn to read the signs of Knowledge, and you will learn to read the signs that the world is giving you. For the world is telling you what is coming. You may not know when it comes or how it comes, but the world is telling you what is coming.

This is redemptive to you, for you have to get serious about your life now. You cannot be lazy, foolish, or self-obsessed. You have to snap out of that. You have to wake up from your nightmare. You have to give up your longing for things that are not taking place. You have to stop condemning others, or holding others to blame for your condition or for the condition of the world, for that is a waste of energy with no good outcome.

Here the Great Waves of change, which seem so overwhelming and so difficult once you have the courage to face them, in reality provide the power of redemption for you. For they require a greater strength and a new kind of relationship with yourself. Doing great things will restore your own value to your life and your own sense of power and ability. Facing great change or great needs will call forth from you a

BUILDING YOUR ARK

deeper and more profound strength than you would ever be able to discover otherwise.

Great times create great people. No one becomes great in the face of complacency or indulgences. Here it will not be just a few exceptional individuals in the human family that will make all the difference, for many more people must be called to their higher purpose now. And higher purpose is not just simply a definition or an idea. It is an immense sense of responsibility and self-determination leading to dramatic action.

Here you may have to give up your refuge in the country or your pleasurable indulgences to a certain degree so that you can respond to the world powerfully. Knowledge will help to move you into the right position, the right situations, to call forth your gifts from you and to give you the greatest opportunity to meet those individuals who are destined to be part of your greater service to the world.

Here you must be both a leader and a follower all at once. You must lead your mind. You must lead your emotions. You must govern your habits. You must override your weak and self-defeating tendencies. Yet you also must follow, for only Knowledge knows the way. And you must follow Knowledge, and yield to Knowledge, and choose with Knowledge whenever real decisions are required.

When people think of great change in the future, they think of what governments must do, what nations must do, what large organizations must do. Yet while nations and large organizations will have important parts to play, it is the power of the individual that ultimately will make the difference.

If people are going to be like sheep, then they will resign themselves to whatever powers are governing them, and they will go wherever these powers want them to go. It is this kind of self-sacrifice and irresponsibility that has led the world to the precipice, that has taken humanity to this dangerous and great threshold.

Therefore, you must think like a person who is being guided by the power of Knowledge. And you must discern Knowledge and distinguish it from your own compulsions and your own adamant beliefs. This takes time. It takes preparation. It takes wise counsel.

You cannot do all this on your own. You will need a preparation you did not invent for yourself. And you will need strong individuals in your life to help clarify your life, to show you when you are in error, and to give you the strength to choose what is really true.

Here the Power of God brings people together in meaningful ways. For alone you can do nothing. But Knowledge will unite you with other people so that your power will be amplified, your strength will be reinforced. This is why relationships based on Knowledge are so powerful and are so resilient, and why they can weather the changing storms of the world, where other relationships will fall away and will collapse.

You must build your ark, and you must support others building their ark, for you need to have strong people around you. In relationships, people are either weakening you or strengthening you. There are no neutral relationships. Anyone with whom you spend time and energy is either strengthening your preparation and awareness or is weakening it. Even lovely, wonderful people can be a great disincentive here. Knowledge will help you to discern who to be with and how to be with them, thus ending endless speculation and debate about who to be in relationship with.

BUILDING YOUR ARK

If people cannot support your building your ark, then they are taking you away from your fundamental responsibilities. If they think you are being foolish, or extreme, or eccentric in having this awareness, then you can be assured that they are weakening you and are weakening your position.

Be careful of rationality because rationality here is usually a reinforcement of the past. People will use rationality to talk themselves out of their own awareness and their own deeper experience. You will appear to be unreasonable to other people. They do not think the Great Waves of change are coming. They think the future will be like the past. They are not responding to the great Calling that is going out in the world. And though they may be very intelligent and intellectual, or very smart in other areas, here they are being blind and foolish.

When the Great Waves come, people are standing on the beach like it is any other day. They think tomorrow will be like yesterday. They think today will be like any other day, not knowing that today will be very different. The animals are heading for the hills. People are lying on the beach. Who is being intelligent now? Who is being foolish now? So do not fall into the trap of thinking you must have a consensus of other people to make the important decisions that are before you. For you do not want to be lying on the beach when the Great Waves come. And you know they are coming.

Here you are governed not so much by fear, but by responsibility, by the ability to respond to your own internal experience and to the signs the world is giving you. When the animals build shelters or seek refuge for the winter, they are doing that because they are being responsible. But humanity demonstrates an incredible degree of irresponsibility in this regard. The oceans are rising, but people are building mansions on the beach. The deserts are drying up, but people are moving there for retirement. Food production is being impacted by the Great Waves of change, but people move out far into the country, far away from the sources of food and food distribution. Governments are becoming weaker and more unstable, but people are still waiting as if they were on welfare of some kind.

Do you see the problem here? Your ancestors long ago had to pay attention to the environment at all times because they had to be aware of the risks and the changing circumstances of that environment. But people living in the wealthy nations, particularly, have lost this essential responsibility to the environment, and this essential relationship with the environment. Now they take for granted that everything will be there for them, provided by other people, provided by complex networks of organizations. It is unthinkable to them the possibility these organizations might not function well, or that there will be breaks and discontinuities in delivering them everything that they need and expect.

The Great Waves of change will create discontinuities. You must be prepared for this. These are the changing times. You are preparing for a world in decline. You are preparing for the future that will be unlike the past. Therefore, you cannot be relying upon assumptions and beliefs, because assumptions and beliefs are all reinforcements from the past.

Even if you are the only person who is thinking of these things, you must honor your deeper experience. You must build your ark even if no one else is doing it. You must engage in the deeper re-evaluation of your life—re-evaluating your possessions, your obligations, your relationships, your activities, where you live, how you live, how you travel about—all in light of the Great Waves of change. The alterations you make to your life here, if they are wise, will give you greater

BUILDING YOUR ARK

stability and will give you greater assurance that you will be able to act wisely in the face of uncertainty and great change.

People who do not do anything will not have this certainty. They will not be resourceful. They will not be able to respond in the moment to the changing circumstances of their lives.

You build these skills through the preparation, by taking the Steps to Knowledge and following the steps that Knowledge will provide. The Creator of all life has given a set of guidelines for preparing for the Great Waves of change, but these are only beginning guidelines, for everyone has a different gift and a different path to follow. People's circumstances are different, and so the guidelines only provide the basics to enable you to begin your preparation.

You may have to turn a hundred times in the future. It is Knowledge that will teach you when and how to turn, how to change, what to decide to do and not to do. That is why Knowledge is your greatest asset and your most powerful endowment.

God saves you through Knowledge. God redeems you through Knowledge. God ends the Separation through Knowledge. This is why the inner preparation is greater than the outer preparation. This is why you must respond to the signs that God is giving you, on the inside and on the outside.

Building your ark is not just to protect you, it is to enable you to give to the world, to find the people and circumstances that will call forth from you your greater gifts, which you yourself could not call forth.

The answer is within you, but the calling is out in the world. You cannot call these gifts from yourself. And your greater purpose is not what you think it is. You may think you are to be an artist or a musician, when in fact you have a greater service to render somewhere else. You might think that your greater purpose is to be what you want it to be, but in almost all cases it is really something else.

Only Knowledge knows, and you can only find what Knowledge knows by following Knowledge so that it may reveal to you the reality of your greater purpose in the world. You build your ark for this, so that you may be a contributor to a world in change and not merely a victim of it.

Your preparation begins right now, today, and tomorrow, and the days to come. There is no thinking about it. You can do that forever without any clear resolution. Allow your deeper response to guide you, and do not listen to the weakness of your thoughts and your emotions.

This is a gift of Love because God loves the world. And God wills that humanity prepare for the Great Waves of change, survive the Great Waves of change and build a greater and more just civilization in the future. But what God wills and what people want and are aware of are not the same. So it is bridging this gap, then, that provides the greatest resource and assurance for humanity, and the greatest meaning and purpose for your life.

As revealed to Marshall Vian Summers, January 11, 2009

CHAPTER 14

Finding Certainty and Strength

Humanity is entering uncertain times. Great Waves of change are coming to the world—generating increasing economic and political instability, disrupting and even destroying the production of food, pushing humanity beyond the limits of what the world can provide.

It is a time of great uncertainty, a time of upheaval, a time when humanity must face the reality that it has overspent its natural inheritance and must now adapt to a changing set of circumstances in the world. It is a time when the risk of competition, conflict and war over the remaining resources will become ever greater—a time of great risk; a time when humanity must choose whether it will fight and compete amongst its nations and tribes or whether it will unite for the preservation of the world. This is a decision not only for leaders of nations; it is a decision that must be made by citizens everywhere.

It is a time of great danger for humanity, but it is also a time of great promise. For it is in the face of the Great Waves of change, and in the face of humanity's encounter with intelligent life in the universe that your greatest opportunity to establish a new foundation and direction for the human family can be realized and fulfilled.

Every person will have to face increasing uncertainty, even amongst the wealthy nations of the world, even amongst the most privileged.

PREPARING FOR THE GREAT WAVES OF CHANGE

They have far more to lose in the face of the Great Waves of change, and they are not adapted to living under more pressing circumstances.

The very poor, however, are at the greatest risk, for they have nothing to fall back on. There will have to be a great provision for the poor peoples of the world.

This will have to be a responsibility that nations carry, and that the whole world carries. For you cannot afford to have nations collapsing, governments collapsing, economies collapsing in the face of the Great Waves of change.

It is as if the bill has come due for decades and centuries of overuse of the world and misuse of the world. Now greater care must be brought to bear, and far greater cooperation. It will change the priorities and the values of people out of necessity. It will either forge a greater union or a greater destruction.

You cannot run and hide from the Great Waves of change or from the presence of foreign, intervening forces who are here in the world trying to influence and take advantage of a weak and divided humanity. It is a time of great reckoning, a time of great decision making, a time when each citizen will have to decide what they will emphasize.

As uncertainty grows around you, as your faith in leaders and institutions becomes shaken and challenged, as wishful thinking and personal plans and hopeful expectations begin to evaporate in the face of growing difficulties, people have a great opportunity to recognize the deeper powers that the Creator of all life has placed within each person to guide them, to protect them and to lead them to a greater contribution in the world.

FINDING CERTAINTY AND STRENGTH

It is indeed in the face of increasing uncertainty that the possibility that people will find this greater power and certainty within themselves becomes heightened. Once you lose faith in things that have little substance or little assurance, then there is a greater possibility you will turn to that which is truly certain—that which is unshaken by the world, that which is not afraid of the world, that which can navigate the difficult and uncertain times ahead.

Here you must learn that you cannot be certain about the outcome of how things will turn out. You cannot be certain of the outcome that humanity's efforts on its own behalf will be fruitful and productive in the way that is hoped for. You cannot be certain that your governments and leaders can navigate the increasing political and economic turmoil that the Great Waves of change will produce.

But you must be certain increasingly that you can navigate these difficult times; that you have a pilot within yourself that can guide you; that you can face whatever life presents to you and not fall apart, or give way, or take desperate actions, or make foolish assumptions.

Here the wealthy people have less adaptive abilities than the poorer people. The wealthy, who have been insulated by their wealth and privileges, will find themselves at a great loss as to what to do. That will be great stress upon them, for they are ill equipped to deal with changing circumstances of this nature.

Therefore, your certainty must not be in the outcome, but in your ability to go through the process of change. And this certainty cannot merely be a hope or a wish. It cannot merely be a faith in oneself. It has to be based on real skill, clarity and objectivity.

You will see around you that people will go from living in hopeful expectation to despair without finding the real middle ground. They

PREPARING FOR THE GREAT WAVES OF CHANGE

will fall from their ideal world and their great assumptions about security and success into despair, hopelessness, desperation. They will go from one extreme to another, from one kind of foolishness to another, from one kind of dangerous attitude to another.

Where people need to be is in the middle: clear, objective, sober, determined, and compassionate with others. For it will take great patience and forbearance to not fall into the temptation to condemn or to hurt others in the face of such disruptions.

You have to be collected, you have to be observant, and you need to prepare. Lack of preparation here is a critical point because people will be taken off guard. They will not see the waves of change coming and will find themselves all of a sudden in a great difficulty without any resources available to help them, really.

In the future, your government will not be able to protect everyone, will not be able to shore up everyone, will not be able to feed and house everyone. In the world, there will be great environmental disruption as arid regions become uninhabitable for large numbers of people, and there will be great migrations, great human need. In the wealthy nations, there will be a loss of wealth. Economies will shrink, people will have to collaborate to provide for themselves and to help their communities.

If you see this coming, and it is coming, then you can begin to re-evaluate your life: how you live, where you live, how you travel about, how many resources you will require to sustain yourself and your family. You will have to see if your employment is sustainable into such a future. You will have to look at your skills, your strengths, your weaknesses. You will have to take a very objective assessment of everything you have, everything you do, everything you own and, perhaps most importantly of all, the quality of your relationships.

FINDING CERTAINTY AND STRENGTH

If you associate with people who are weak and prone to foolishness or foolish assumptions, well, they will be of no help to you. In fact, they will be a great problem for you.

God knows that you were going to enter such a world at this time, and that is why God has given you the power of Knowledge—a deeper Intelligence, an Intelligence that functions beyond the realm and the range of the intellect, an Intelligence that is not afraid of the world, that is not governed by people's political or religious beliefs, conflicts and perspectives.

It is like a pilot taking you through a dangerous channel, but you must be really ready for this passage. You cannot sit idly back and think that God is going to take care of everything for you. For God has sent you into the world to take care of everything—you and everyone else.

But God knows you cannot do it alone, and God knows the world is a very corrupting and diluting environment. And so you have been given the greatest endowment possible. But incredibly very few people are aware of this, the power and presence of Knowledge within themselves. It is the source of your certainty, self-confidence and true abilities.

Some people think when they learn of the Great Waves of change and are able to experience these things themselves, they think they must stockpile food, they must build a fortress, they must create a defensible position, against everyone and everything.

But this is not guided by Knowledge. This is a fool's response because you cannot stockpile food for a lifetime. And if you stockpile too much, people will come and take it away from you. You cannot live in fear and trepidation if you are to be of service to the world, which is really your purpose for coming here.

Knowledge will protect you, but it will not imprison you. It will take you to a position where you can be a real contributor to the world and to other people, and to be a beneficiary of change rather than its victim.

Knowledge is not here to make you rich, or to have you barricade yourself against the world, or to run off into the wilderness and try to survive. That is what the mind thinks, your personal, social mind. It thinks of such things. But this is foolishness.

You cannot run away and hide from the Great Waves of change. And if you try to barricade yourself, you will just become isolated and in danger. You cannot take up arms against your neighbors because that reduces everyone's ability to survive and to adapt.

You need the power of Knowledge now to guide you because your intellect cannot do it. People claim they have solutions to the Great Waves of change, it is just a matter of political will and policy, or it is just a matter of supporting new technology. But people who think these things do not understand what is really coming. They do not realize that it will take a thousand solutions to mitigate the impact of the Great Waves of change, and it will take years to accomplish this. They do not realize they will have to adapt significantly to the changing circumstances of the world.

They think that they can create a defense at the level of policy and technology. They underestimate greatly what they are really facing here. And they do not see that humanity will not prepare itself adequately to meet the great challenges that are coming now.

FINDING CERTAINTY AND STRENGTH

But this is not a cause for hopelessness because people still have to become aware and to prepare. And the degree to which they do will determine the quality of the outcome for them and their ability to undergo a process of great upheaval and change.

Humanity is not doomed, but it will be extremely challenged. It is facing this challenge, understanding this challenge and finding the resources in oneself to undergo the process of preparation and change that really are important. And the value is not merely survival; the value is that it will elevate people into a greater set of abilities and a greater service to others.

The kind of world you will have in the future and the kind of world your children will inherit will be dependent upon the degree to which this recognition of preparation and contribution can take place.

Therefore, do not be certain about the outcome, for that is self-deceiving. Do not be certain that governments or technology will be able to meet all the challenges, for that will blind you to your own need to prepare and it will blind you to the power and the impact of the Great Waves of change.

What you must be certain of then is your ability to face, to comprehend and to undergo this process of change and upheaval. And this cannot be a wish or a hope. It must be based on a real foundation within yourself. Preparing your circumstances will be very helpful, but most of the preparation here is internal.

People have become so reliant upon other people to guide them, to lead them, and upon the seeming powers of technology, they have lost their self-reliance. They have lost their intuition. They have lost their ability to respond to unprecedented and unexpected events.

PREPARING FOR THE GREAT WAVES OF CHANGE

They have lost what nature has provided that assures and supports your survivability and your adaptability.

Sixty percent of the preparation will be internal; [forty] percent will be external. The more you prepare, the more competent you will feel you are. If you do nothing, you will have no confidence. If you wait for other people to fix the problem for you, having certainty that they can do this and that everything will turn out fine, you will have no courage, you will have no strength, you will have no competence to face the Great Waves of change.

The process begins by awareness. It might be frightening at first, but you want to move beyond your emotions into a place of greater objectivity. When you realize that you cannot run and hide and no one is going to save you, then you can enter a state of mind that is much more clear and responsive.

Here you set aside self-obsession, unnecessary hobbies, your plans and your goals. You start to look at where you are losing energy to people and to involvements that have no real significance in the face of such great change.

You look at your community and you see that the elderly are vulnerable; children are vulnerable. And you see the magnitude of the need for communities to prepare, for people to become aware, for people to undergo the deep evaluation of their lives that you are now beginning to undergo. If people are not ready, they will be overtaken, and then there will be very little for them to do.

In life, you live for the moment and you prepare for the future. You do not just live for the moment. Perhaps if you are an ant, you live for the moment. But even the ant is building a colony, supporting an infrastructure that can withstand immense change.

FINDING CERTAINTY AND STRENGTH

This is where you wake up from your dream of personal fulfillment, or your dreams of romance or your dreams of your own involvements or obsessions, to wake up to the world. The world is knocking at your door, saying, "Wake up, it is time to wake up now. It is time to begin to look at your environment, to look at your life, to look at what you are doing. The Great Waves are coming over the horizon."

This is not a matter of belief. It is a matter of recognition. It is not a matter of perspective because no matter how you look at it, the Great Waves are coming, and some are already here. It is not a matter of being loving or fearful. It is a question of whether you can see or not see, whether you can respond or not respond, whether you can prepare or not prepare.

This is a great wake-up call for the human family, a wake-up call telling you that you live in an environment, and the environment has been changed, has been altered, has been despoiled in many cases, and now you must face the consequences of this. This will utterly change your priorities, your objectives, your emphasis.

Instead of just being a locust upon the world, consuming everything in sight, you will become really concerned about the welfare of your town or your community, or if your city will be able to undergo this kind of strain and challenge. You will be concerned about the welfare of people who are vulnerable, who are elderly, or very young, or disabled. You will look to your family; you will look to your network of friends and associates. You will look to the survivability of your business. You will begin to assess what is strong, what is weak, what can last, what will not last. The more you do this, the greater will be your self-confidence, and the stronger will be your experience of Knowledge.

PREPARING FOR THE GREAT WAVES OF CHANGE

You cannot freeze or be paralyzed, you cannot run away and hide, or go around blaming other people, or be hysterical, or break down in self-pity. You must gain hold of yourself, find your center, your strength, get over your initial reactions. You are on a ship that is slowly sinking. What are you going to do about it? Run around screaming and yelling, or be enraged at other people?

For everyone has created this condition, you see. There is really no one to blame. There are only people to engage with. You want to be part of the solution and not the problem.

Humanity will have to employ all of its creative talents in science, in social progress, in political development, humanitarian aid, international relations—everything, as if your whole world was at stake.

You now face competition from exploratory groups from the universe who will seek to take advantage of this situation. They are not military powers, but they carry great strength and persuasion. They will attempt to divide and conquer humanity through the power of this persuasion and through promises of technology by claiming that they live in peace and can rescue humanity from its predicament.

But it is all a deception, for they cannot save humanity. They can only take advantage of humanity. And this is the perfect environment for them to do this. As people lose confidence in governments and institutions, they will turn to other powers to rescue them.

Such powers are already in the world, claiming to be here to save humanity from itself. It is a more intelligent way to conquer. It is a more brilliant plan. They will not use force because force destroys the value of the world, and they need the human family to work for them.

FINDING CERTAINTY AND STRENGTH

So destruction is not their aim. They cannot live in your world, so they need you and your infrastructure that you have built.

You must see that all of this is not overwhelming. It is really nature. It is really natural forces. Competition is part of nature. You now have competition from the universe around you, from the Greater Community of life in which you have always lived, and which you must now contend with.

It is the overwhelming nature of the great change that will call you out of your fixations, your beliefs, your perspective. Only something this strong and this demanding can really produce a real shift in how people think and their ability to see and to recognize their situation.

Humanity could lose everything, everything it has ever created for its benefit, in the face of the Great Waves of change, and in the face of Intervention from the universe. But to see this you must have clear eyes. You cannot be in denial or avoidance. You cannot be so reasonable that you cannot see something that is unprecedented. You must look over the horizon and see what is coming. It is not a matter of what you want. It is not a matter of what seems reasonable because the future will not be like the past.

Humanity has a greater promise. You will have to build a new future because you cannot continue what you are doing now. But what will create the incentive to do this, the commitment to do this, the courage and the sacrifice to do this? Only the Great Waves of change and humanity's facing competition from the universe will really have the power to unite the human family in a very functional way.

For everyone is in the same boat now. It does not matter whether you are rich or poor, whether you live in this country or that country. You are all facing the same outcome.

PREPARING FOR THE GREAT WAVES OF CHANGE

If human civilization collapses in the face of the Great Waves of change, other powers will come here to take over. Some of those powers are already in the world—planting the seeds of dissension, preying upon human weakness and human superstition, encouraging people to believe in the Intervention and to lose all hope and confidence in human leadership, to weave a web of conspiracies that will make people fear and doubt one another so much that they will never be able to trust each other enough to collaborate.

The Creator of all life has sent a New Message into the world to reveal the power and the reality of the Great Waves of change and to reveal to humanity the hidden presence that is in the world today that is seeking to undermine human authority and human sovereignty in this world.

This New Message is here to speak to the power and presence of Knowledge within each person and each person's greater responsibility to be of service to a world in need at this great turning point in humanity's evolution.

The New Message presents what humanity cannot see and will not see. It gives a clear picture. It is not a product of any existing world religion or institution. It is a gift of love and concern and confidence from the Creator of all life. It teaches that all religions were inspired by God. It teaches the power and the presence of Knowledge, the essence of human spirituality, and that each person was sent here to give a greater gift to the world, and that human freedom and human cooperation are what will allow these great gifts to emanate within the individual and to find their fullest expression in the circumstances of life.

The New Message teaches you where certainty can be found and where it cannot be found. It will teach you the difference between real

FINDING CERTAINTY AND STRENGTH

faith and faith that is being misused and misapplied. It will teach you to value your life and the lives of others, and to learn to hear the great need for Knowledge in the world, and to listen to the evidence of Knowledge in others.

Humanity will have to become strong and united. It will have to restrain its violent tendencies. It will have to overcome its historical conflicts and animosities. Nations will have to cooperate if they want to survive the Great Waves of change.

But what can give humanity the power to overcome all of these extremely powerful tendencies—the tendencies towards conflict and conquest, towards hatred and destruction, towards greed and corruption? It cannot merely be a set of ideals or an ethical proposition. It has to be something that is so dramatic and so potent and so real, that you are faced with a fundamental choice: unite and survive, be divided and you will fail.

You are now facing such a great challenge. You are not certain of the outcome. You are not certain that things will work out fine. You have never been through this before. The past cannot now tell you what the future will be like.

Tribes and nations have collapsed in the past, but never has the entire human family been facing a world in decline, a world of depleting resources, a world of environmental degradation and change. Never have you had to face competition from the universe. If you cannot consider these things, then you will not see the need to prepare, and you will not prepare. You will just hold onto your old values, old notions and ideas, and will be blind to the signs of the world, and will be deaf to the voice of Knowledge within yourself.

PREPARING FOR THE GREAT WAVES OF CHANGE

Humanity is entirely capable of failing and has shown all the right tendencies for failure. That is why this is an internal matter, a matter of conscience, a matter dealing with the deepest conscience within you, which is Knowledge, which is the conscience that God has placed there, not a conscience that your society has tried to cultivate.

An unprecedented set of circumstances requires an unprecedented response. Does humanity have this responsibility, this ability to respond? You cannot answer that for others, but you can answer that for yourself. And that is why this is a challenge to you, first and foremost.

There is no sitting by in the face of the Great Waves of change. There is no deferring to others. You must look to your life, to your circumstances, to your family and to your relationships. This is how you begin.

You do not neglect these things and try to change the world. You must begin here. You must have a more secure position, so you do not end up being a victim when the Great Waves begin to strike. You must be able to be strong enough to weather the shocks and to use the guidelines that the New Message has provided you that will enable you to begin your preparation.

This is a Message of love and compassion. This is God warning you, blessing you and giving you a preparation. This is alerting you to the reality of the future, the future that will overtake the present as it approaches. This is to give you time to see, to know and to prepare, and to undergo a deep re-evaluation of your life.

This is to show you where your real certainty resides and to teach you the difference between the Power of God and all the false Gods that you have created, believing that they will guide and protect you.

FINDING CERTAINTY AND STRENGTH

This is to encourage you to support human unity and cooperation, no matter what the odds, no matter what the corruption, no matter what the deception that has gone on before. For what other choice do you have? You need everyone now to work, and Knowledge will give you the strength to do this because it is not concerned with the odds.

It is here to give, it is here to contribute, it is here to strengthen people, to unite people, to encourage people. It is the antidote to fear. It is the antidote to evil. It is the antidote to self-deception, and it is the antidote to self-defeat.

Listen with your heart and you will know. Listen with your mind and you will be confused and take issue.

Not everyone has to respond, but enough people must respond or the tide will turn against the human family, and the danger and the requirements will seem overwhelming. It was not raining when the ark was built. Do not wait for the rains to come, or there will be no building of an ark.

You will need great courage. You will need real determination. You will need to persevere. Your ideas and beliefs cannot give you this courage, this perseverance, or this strength. It must come from deeper within you, and that is why Knowledge is the key to whether people will see, know and act, according to the demands of the world.

You begin here and you proceed step by step. You do not know the outcome. You cannot assure the outcome. There are no guarantees. There is only contribution. There is only moving in a right direction. There is only choosing strength over weakness, certainty over illusion, compassion over hatred.

As revealed to Marshall Vian Summers, January 13, 2009

CHAPTER 15

THE GREATER MIND OF KNOWLEDGE

There are two great events that humanity is now facing, events greater than anything that humanity as a whole has ever encountered before.

You are facing the Great Waves of change, brought about by environmental disruption and misuse of the world's resources; violent weather; environmental degradation; diminishing energy, food and water resources—great environmental changes that will make life in this world more difficult, literally for everyone, including people in the wealthy nations. This is just beginning to be well documented in your world today, but its impact will be great and far reaching and will literally change the landscape of the world so that the 21st century will be very different from the 20th century.

The other great challenge and change facing the human family is your encounter with intelligent life from the Greater Community, from the universe around you. Very few people in the world today are aware that contact is occurring, and the vast majority of those few who are, do not understand it correctly. This contact has the power to deny humanity its freedom in the future, to cast human civilization under foreign control and domination.

It is not conquest in the usual sense of someone coming and violently overtaking you, for those who seek to gain control of the world and the world's peoples are small in number and do not have military assets. That is not their strength. They are cunning and intelligent. They realize they could gain control of this world through non-military means, and they are seeking to do so even at this moment.

There are very few people in the world today who are aware of the Great Waves of change that are coming, and what they really signify, and the kind of impact and consequences they could create in the world, both now and in the future. There are very few people in the world today who realize that an Intervention is occurring from beyond the world and that it represents a competition for power here—for power, influence and future control.

Together, these things will indeed be overwhelming if you can face them. Indeed, there are some people who are beginning to be aware of them but cannot face them because they are overwhelming, because they will require you to change, and to re-evaluate your circumstances and your situation, and where you stand in your position with life itself.

So there is ignorance and there is human denial. But the Great Waves of change are coming, and the Intervention is already occurring. These are two great realities that you cannot change, but the outcome is largely in the hands of human beings and human communities.

The first great challenge is to face the great challenges that are coming into your life. Though they may seem overwhelming and terrifying to your mind—to your thinking mind—the deeper Knowledge that God has placed within you is well prepared for them. Indeed, seen from a greater perspective, this is why you have come into the world—to serve humanity and to contribute to a world in need, facing these two very great challenges.

THE GREATER MIND OF KNOWLEDGE

You must have a greater perspective to see and understand this, a greater sense of yourself, a greater recognition that you have come into the world for a greater purpose—not merely to survive, not merely to consume, not merely to gratify yourself, but to serve humanity at a time of its greatest need. To go from your current understanding to this greater understanding represents the purpose and the necessity of your spiritual growth and awakening.

To shift from purely a human perspective of one who is afraid of death, afraid of loss, who is trying to acquire as much as possible for themselves to this greater understanding that you have come into the world for a higher purpose represents the great transition that you must undertake if you are to be able to navigate the difficult times ahead and to fulfill your mission here in the world.

Your purpose here was established before you came into the world, but how it will be expressed will be determined by so many things, so many circumstances. Certainly, your own decisions are a big part of this, but what other people will choose will also have an impact on when and how and even if you can fulfill your destiny here.

There are certain individuals you will need to meet who hold some of the pieces to the puzzle of your life. Their arrival and participation in your life, if correctly seen and correctly understood, will enable you to make this great transition and will give you the strength, the encouragement and the companionship that you will need to be a real contributor in a radically changing world.

Here you must not be concerned about what everyone else is doing. Here you cannot allow yourself to lose heart and to become discouraged by the lack of awareness and response of people around you, for you have your own journey to take. You have your own calling to respond to. You have your own transition to undergo within your own life and with your own awareness.

God knew what you would be facing before you came into the world. God knows what is coming on the horizon of the world, even now. And God knows that you could not possibly recognize all these things and prepare correctly and accordingly without God's Wisdom to assist you and without the Power of God to give you the courage and the commitment to prepare yourself for the Great Waves of change and for the Greater Community in such a way that you will be able to be a real contributor to the world and to the future and the freedom of humanity.

God knows that your intellect alone cannot comprehend these greater movements of life. God knows that you would be far too affected by your social circumstances and the influence of other people to have much of a chance to be able to see these things clearly and to respond accordingly.

Therefore, God has given you part of God's Wisdom, Intelligence and Power. You carry within you, then, a deeper Mind, a Mind beneath the mind, a Mind that God has placed there.

In reality, this is who you really are, this greater Mind of Knowledge that God has given you. That is the Mind you had before you came here, and it will be the Mind you have after you leave this place. It is alive within you now. And it is within you to guide you and to protect you, to help you make the corrections you need to make in your life, to build a foundation for the world, a foundation for yourself, and to assist and help in building a foundation for others so that humanity may have a future, a greater future than its past.

THE GREATER MIND OF KNOWLEDGE

Without this Knowledge, you would not be able to rise above the influence of the mental environment around you. It would be unlikely that you would be able to overcome your social conditioning and your religious prejudices and your personal attitudes sufficiently to be able to live a greater life and to provide a greater service in the world, a service you were sent here to perform.

That is why God gave you Knowledge. It is not an intelligence you can manipulate. It is not something you can control or dominate. It is far more intelligent than your intellect. It is without fear. It is without confusion. It is without hostility. It does not need to compare and contrast and speculate. It does not need to dominate and control. It does not need to destroy others so that you may feel powerful.

In fact, it does not demonstrate most of the qualities of your intellect. This is because your intellect is a product of the world, and Knowledge is a creation of God. Your intellect reflects the world and your human-animal nature, but Knowledge reflects the Mind of God and your Divine nature. One is temporary. The other is permanent.

Ultimately, in the true hierarchy of your Being, your intellect is meant to serve Knowledge. However, at the beginning, when most people realize they need to reclaim their true identity and their greater power and begin to take the Steps to Knowledge, they want Knowledge to serve their intellect. They want Knowledge to answer their questions and to allay their anxiety. They want Knowledge to give them things they want. They want Knowledge to serve them as if Knowledge were a servant of the intellect. But Knowledge is not the servant of the intellect.

Yet Knowledge is here to serve you and to serve the world. It is not afraid of the Great Waves of change. It is not afraid of Intervention from the Greater Community. In fact, it has been waiting for them, prepared for them.

How different this is from your intellect, your personal mind, which is terrified of the world, which is afraid of loss and deprivation, which is full of anger and rage against others, which can never seem to gain control of events around you sufficiently to provide any real security for you. The contrast between your intellect and Knowledge is so great, in fact, that you cannot even compare them. Yet this will seem mysterious to you until you can begin to experience Knowledge itself.

Your initial experiences of Knowledge will be at moments a sense of urgency or a sense of need to change; the recognition you must leave something; the recognition you must find something; an awareness of a future event that takes place; a moment of compassion and selflessness. All of these represent the evidence of Knowledge, which is still latent within you. It has not arisen in your awareness sufficiently that you can really experience its power and its efficacy.

That is why you take the Steps to Knowledge, because you yourself need it desperately. All of the certainty and security you were trying to establish in your relationships, in your work, in your personal affairs, in your intellectual understanding—these can only be provided by Knowledge.

Without Knowledge, you are cast adrift in a world of dramatic change—a beautiful world, but a dangerous world as well. And all you can hope to do is adapt and to survive with as good an attitude as possible. But this is not your destiny in being here, to simply survive for a few years with as much courage and stability as you can muster. That is not why you were sent here.

THE GREATER MIND OF KNOWLEDGE

If you knew your true nature, this would be as clear as day. But Knowledge has not yet arisen in your mind sufficiently for you to have this awareness, for you to have the stability and the security and the power that Knowledge will give to you.

Therefore, the most critical thing now is for you to realize your need for Knowledge—the recognition that you must find what you came here to do and that you cannot be wasting your time, your days, your months, your years with all of your restless searching and your fruitless pursuits, caught up in a web of desire and denial that leads you nowhere, that keeps you in a state of poverty.

Therefore, you must feel the need for Knowledge and also to recognize that you have experienced it, however briefly in your past, and that you need this certainty and that you need this greater experience to know what to do. For in the face of the Great Waves of change, you will not know what to do. In the face of the Greater Community and the Intervention—should you become aware of it, should you have the courage to face it—you will not know what to do. You will feel hopeless and helpless in the face of these two great events.

But Knowledge within you is not hopeless, and it is not helpless. It will begin to reset your life and to build strength where it is needed and to reorient your awareness to greater things that are happening in the world and to the greater movement of life that is occurring, even within you.

Therefore, at the beginning, it is recognizing the need for Knowledge and then taking the Steps to Knowledge. The New Message that God has sent into the world, the New Message for humanity, provides the Steps to Knowledge. There are other pathways to take, but nothing as clear as this.

PREPARING FOR THE GREAT WAVES OF CHANGE

It is a gift of God. It is not a human invention. It is not a process that anyone made up. It does not represent an eclectic approach. It is something new, powerful and original. And it is here without any corruption. It has not been altered or manipulated, used by religious institutions or governments. It is pure—the Steps to Knowledge.

You need to take these Steps now, for your life is important, but you do not have much time. Time is of the essence. Every day is important in determining whether you will prepare for the future or not, in determining your survivability in the future and the freedom of your mind. You do not have all the time in the world to think about it, for that time is wasted, and time is of the essence.

Regarding the Great Waves of change and humanity's recognition that Intervention is occurring from races from the Greater Community, humanity is very late in recognizing these things. There are very few people who can see any of this clearly at all. This has put the entire human family and civilization into great jeopardy, greater than you now realize.

Therefore, the great need is not only your personal need for certainty, purpose, meaning, and direction. It is the need of humanity. You alone will not be able to meet the need of humanity, of course. But you are destined to play an important part in determining whether humanity will have the clarity and the strength and the will to face the great change that is coming and to [unite] in the face of the Greater Community to preserve human freedom and sovereignty within this world—a freedom and sovereignty that are now being challenged as never before.

You are meant to play a part in this. You cannot create that part. You cannot see that part at this moment, for you have not gotten that far up the mountain yet. But you need to take the journey of preparation

THE GREATER MIND OF KNOWLEDGE

that will take you there. For the only way to get up this mountain is to climb it, to take the steps. Nothing is going to lift you up to a vantage point where you can see the world clearly. You have to make this journey. There are no exceptions to this.

And you can see why this is the case because you must gain the strength, the commitment and the confidence to gain the power of Knowledge. It is not something that simply happens to you as if you were struck with a bolt of lightning. Even if the Angels appeared to you and told you what you must do, at this moment you do not have the strength to do it. You do not have the trust to do it. Your life is in disarray. You would have to alter your circumstances. You would have to rearrange your commitments and your obligations.

In other words, you would have to take the Steps to Knowledge. There is no getting around this. From where you are at this moment, you cannot assume a greater life. And you will not have the power of Knowledge to protect you and to guide you because you are not aware of it sufficiently, and you do not trust it yet sufficiently.

That is why you must take the Steps to Knowledge. You do not have decades to do this. You do not have all the time in the world to be ambivalent in the face of this great need and challenge.

Within you, you know this because you are uncomfortable. You are restless because there are things you must do that you are not doing. And there are things you are doing now that you need to stop doing. And you are restless. Your soul is restless because at the level of your soul, there is only being in the world and completing your mission. Your mind may want an endless list of things—pleasures and comforts and assurances and opportunities. But to your soul, it is all about fulfilling your mission here.

Knowledge, then, is about fulfilling the need of the soul, for this is the most profound need within you. And as this need continues to go unmet, you are anxious, you are restless, you are upset, you are uncomfortable, you are agitated because at the very core of you, the great need is not being recognized and fulfilled.

How could you have any equanimity if the need of the soul remains unrecognized and unfulfilled? How could your life have any stability? How could your thinking be coherent without this need being recognized and fulfilled?

Most people are slaves to their social conditioning, to their fears, to their desires, to their religious prejudices. They cannot even think for themselves.

Knowledge is the salvation, for only Knowledge has the power to liberate you from all of these things—from a bondage that is inevitable as a result of being in the world.

That is why you must take the Steps to Knowledge. And God's New Revelation for humanity provides the Steps to Knowledge. It is not the only pathway to Knowledge, but it is the pathway that God has given to prepare you for the Great Waves of change and for the reality of the Greater Community, and to prepare humanity for its future within a Greater Community of intelligent life.

It is here that your inner life and your outer life really make the connection. The need of the soul and the events of the world are not disconnected. For indeed it is the world that will call out of you your greater purpose. It is recognizing where you are meant to be and what you are meant to serve and contribute to that will give so much definition and clarity and coherence to your greater work here.

THE GREATER MIND OF KNOWLEDGE

Very, very few people have found their greater work in the world. So do not think that by becoming an artist or a poet or a musician or some other wonderful activity that that will satisfy why you have come here. Those are popular notions with some people, but really your higher purpose is something else. You cannot define it. It is not grand and glorious. You will not be some hero or saint leading the masses to salvation. That is fantasy material. I am talking about reality. We are talking about reality.

Knowledge is the great power within you. It is seeking even at this moment to gain your attention. You are not paying attention, and that is why you are restless and disconcerted.

Learning, then, to pay attention to Knowledge—developing stillness so that you can experience Knowledge, redirecting your mind and your concentration towards Knowledge, developing discernment and discretion in being in the world, refraining from attacking others, learning to comprehend Knowledge in others, recognizing those who are meant to be your true relationships from others who merely stimulate you—these represent only some of the great achievements that Knowledge will provide for you.

As We have said, Knowledge is not a human creation. It is a Divine Creation. It is through Knowledge that God will help you. You may pray and pray and pray for safety and for advantage, for protection for yourself or your loved ones, but it is through Knowledge that God can really help you because you need, even at this moment, the Wisdom of God as it relates to your life and your circumstances. You need the Wisdom of God as it relates to what you need to do and to be, and to the change you must bring about in your own affairs in the world to make it possible to move towards a greater life.

PREPARING FOR THE GREAT WAVES OF CHANGE

This is how God is going to help you. God of the entire universe is not preoccupied with human affairs. God is not following you around like a handmaiden, like a servant, like some kind of errand boy to fulfill your wandering desires.

God has placed Wisdom within you to guide and to protect you and to give you the strength and the wisdom to find your way so that you may assume a greater role in life and provide a greater service to humanity.

People often ask: "Well, is this about intuition?" And We say intuition is not great enough to encompass what Knowledge is and what Knowledge can do through you. Intuition is fleeting experiences of inner certainty and sometimes precognition of events, but Knowledge is so much greater. You really cannot compare them adequately. Perhaps you could say that intuition is the early fleeting experiences of Knowledge. But what Knowledge is meant to be in your life is far greater. For it is a presence. It is a power. It is an awareness. It is a capacity. It is a love at a very different level than what people think of as love. It defies human definition.

If you try to understand it with your intellect, it will always be beyond your comprehension. If you try to make it concrete and simple, it will escape your experience. If you try to grab onto it and use it for your own personal advantage, it will slip through your fingers.

For you were meant to serve Knowledge, for Knowledge is who you are and why you are here. But that is not how you think of yourself at this moment. This is not the awareness that you have yet, so you must take the Steps to Knowledge to prepare yourself for a future that will be unlike the past, to prepare yourself for the Great Waves of change, to prepare yourself to deal with the Intervention from the Greater

THE GREATER MIND OF KNOWLEDGE

Community, to prepare yourself to find those individuals who are destined to help you and to assist you.

You need Knowledge just to be safe in a world that will be far more dangerous in the future. You need Knowledge to have the certainty to live with strength and purpose under very uncertain circumstances. You need Knowledge to rise above the growing conflict and anxiety that will be occurring all around you as people become overwhelmed by the Great Waves of change.

Really, who you are and what you can do must be revealed to you. You cannot figure it out with your ideas. For your life is mysterious, is it not? Beneath the struggle of your mind to understand and to comprehend and to control and to determine events, underneath that surface and awareness and intelligence, there is a greater mystery of your life—who you are, where you have come from, what you are meant to do, where you will go beyond this world. This is the realm of Knowledge.

Save yourself a great deal of time and foolish attempts by not trying to understand it intellectually. It is an experience you must have. It is an experience you have already had, in little ways.

That is why you are searching for purpose and meaning in your relationships, in your work, in your activities, because you are feeling the need of the soul. You are not satisfied with little things. You need more things. You need greater things. You need a breakthrough. And as the world grows more difficult in the face of the Great Waves of change, you need strength and certainty and courage—all coming from the need of the soul, you see? Not merely a psychological need, it is a need of the soul.

PREPARING FOR THE GREAT WAVES OF CHANGE

Therefore, recognize what a great revelation this is for you—that God has given you the power and the strength of Knowledge, hidden deeply within you where you yourself cannot find it. And God has given you the Steps to Knowledge so that Knowledge can emerge within you naturally and appropriately and that you can build the strength to be a vehicle for Knowledge, to build a foundation in your life for Knowledge, to surround yourself with people who can support the emergence of Knowledge within you.

You can perhaps begin to see the need and the problem, but the solution must be revealed to you. For you must understand that God is not going to come and rescue humanity in the 11th hour from the Great Waves of change or from the Intervention that is occurring in the world today. No.

God has placed Knowledge within every person in the world. If even a small number of these people can gain the strength of Knowledge, they will have the power to turn the course of humanity and to provide for humanity, and you are meant to be amongst this number. That is why you cannot be too concerned with others who will not see and cannot see, who will not know and cannot know. You must attend to your own experience and preparation.

That is the calling for you now. That is the next step for you now. Accept that you do not have an answer regarding who you are and why you are here. Accept you do not have an answer about how you are going to deal with the Great Waves of change or with the Intervention. If you have the humility to do this, then you can begin to put yourself in a position to receive what God has intended for you.

Begin to take the Steps to Knowledge, to prepare your mind and your life so that Knowledge can emerge within you, which it will

THE GREATER MIND OF KNOWLEDGE

do naturally. Once you create the environment for it and the desire for it and the capacity for it, it will arise naturally.

It has been trying to arise within you all along, but your mind is somewhere else, your concerns are somewhere else, your energy is being devoted somewhere else. You are preoccupied. You are self-obsessed. You are governed by your social conditioning. And so all the messages that Knowledge is sending you every day are missed. You are not hearing the cues. You are not seeing the signs. You are not responding.

Therefore, to begin to make this great transition to being an inner-directed person instead of being an outer-directed person requires that you take the Steps to Knowledge, and this takes time. This does not happen overnight or within a month or two. This takes time, for it is a journey.

You cannot run up this mountain. This is not some kind of frolic. It is a big mountain, and you must make the journey. And as you make the journey, well, you begin to set aside things that are not necessary or that are weighing you down or that are too heavy to carry.

As you proceed, others join you in your journey, and you realize that you are not traveling alone and that you need their support and encouragement, for you are still down on the lower parts of the mountain, surrounded by the forest. You cannot see where you are. You cannot see where you are going. You cannot see your goal.

You are acting on faith, and real faith is based on the power of Knowledge encouraging you. It is a conviction that you must proceed; you must continue. At the moment, you will not know what you are doing, you will not know why you are doing it, it will seem to be a crazy endeavor, but it is absolutely the right thing for you.

PREPARING FOR THE GREAT WAVES OF CHANGE

This is the journey. This is where you build the inner capacity for Knowledge. This is where you acquire the wisdom you will need to be able to carry Knowledge and to express what Knowledge has given you to provide for others. This is where you mature as a person, and you gain your greater strength and your greater abilities.

Therefore, you begin by recognizing the need and then you take the journey of preparation. It is the journey of preparation that will begin to shift your life and to renew you and to strengthen you and to give you the awareness that your life is greater than what you experience in this world and that greatness is with you now and that you have a destiny in being here, a destiny that perhaps you have only rarely experienced, if at all.

Then you take this journey, and it begins to reveal things to you along the way. You begin to see things and feel things and know things that other people are not feeling, seeing and knowing. You begin to recognize certain things you never saw before, but which were always there.

You begin to hear the need of the soul in others. Instead of just criticizing them or condemning them outright, you begin to hear the need of the soul within them.

You begin to have the courage and the need to face the Great Waves of change instead of denying them or running away from them. And you begin to realize that humanity is standing at the threshold of space, and it is not alone in the universe or even within its own world.

As you make progress up this mountain, taking the Steps to Knowledge, you begin to have personal revelations. As your mind is less preoccupied with its hopeless pursuits and its unhealthy relationships, it frees up tremendous energy for you. And with

THE GREATER MIND OF KNOWLEDGE

this liberated energy, you regain your strength. You regain your composure. You regain competence and confidence. And you gain the strength to make the journey—strength you were not sure you had at the outset, but strength you are now experiencing.

For every step you take up this mountain will give you greater clarity, greater certainty and will break the hold of weakness upon you. You will become stronger within yourself and more certain, but always recognizing the power of the Mystery within you—which is feeding you now, which is encouraging you now, which is giving you strength and perseverance, confidence and patience.

Therefore, you will become strong and humble all at once, for you recognize that the greater power is not your own and that it is feeding you and serving you as you are learning to serve the greater purpose that has brought you into the world.

As revealed to Marshall Vian Summers, October 1, 2007

Important Terms

The New Message reveals that our world stands at the greatest threshold in the history and evolution of humanity. At this threshold, God has spoken again, revealing the great change that is coming to the world and our destiny within the Greater Community of life beyond our world, for which we are unaware and unprepared.

Here the Revelation redefines certain familiar terms, but within a greater context and introduces other terms that are new to the human family. It is important to understand these terms when reading the texts of the New Message and hearing the Voice of Revelation.

GOD is revealed in the New Message as the Source and Creator of all life and of countless races in the universe. Here the Greater Reality of God is unveiled in the expanded context of life in this world and all life in the universe. This greater context redefines the meaning of our understanding of God and of God's Power and Presence in our lives. The New Message states that to understand what God is doing in our world, we must understand what God is doing in the entire universe. This is now being revealed for the first time through a New Message. In the New Message, God is not a divine entity, personage or a singular awareness, but instead a pervasive force and presence that permeates all life and is moving all life in the universe towards a state of unity. God speaks to the deepest part of each person through the power of Knowledge that lives within them.

THE SEPARATION is the ongoing state and condition of being separate from God. The Separation began when part of Creation willed to have the freedom to be apart from God, to live in a state of Separation. As a result, God created our evolving world and the expanding universe as a place for the separated to live in countless forms and places. Before the Separation, all life was in a timeless state of pure union. It is to this original state of union with God that all those living in Separation are ultimately called to return—through relationship, service and the discovery of Knowledge. It is God's Mission in our world and throughout the universe to reclaim the separated through Knowledge, which is the part of each individual still connected to God.

KNOWLEDGE is the deeper spiritual Intelligence within each person, waiting to be discovered. Knowledge represents the eternal part of us that has never left God. The New Message speaks of Knowledge as the great hope for humanity, an inner power at the heart of each person that God's New Message is here to reveal and to call forth. This deeper spiritual Intelligence exists beyond our thinking mind and the boundaries of our intellect. It alone has the power to guide each of us to our higher purpose and destined relationships in life. The New Message teaches extensively about the reality and experience of Knowledge.

THE ANGELIC ASSEMBLY is the presence of God's Angels who have been assigned to watch over our world and the evolution of humanity. This Assembly is part of the hierarchy established by God to support the redemption and return of all those living in Separation in the physical reality. Every world where sentient life exists is watched over by an Angelic Assembly. The Assembly overseeing our world is now translating the Will of God for our time into human language and understanding, which is now being revealed through the New Message. The term Angelic Assembly is synonymous with the terms Angelic Presence and Angelic Host in the text of the New Message.

IMPORTANT TERMS

THE NEW MESSAGE is a communication from God to people of all nations and religions. It represents the next stage of God's progressive Revelation for the human family and comes in response to the great challenges and needs of humanity today. The New Message is over 9000 pages in length and is the largest Revelation ever given to the world, given now to a literate world of global communication and growing global awareness. The New Message is not an offshoot or reformation of any past tradition and is not given for one tribe, nation or group alone. It is God's New Message for the whole world, which is now facing Great Waves of environmental, social and political change and the new threshold of emerging into a Greater Community of intelligent life in the universe.

THE VOICE OF REVELATION is the united voice of the Angelic Assembly delivering God's Message through a Messenger sent into the world for this task. Here the Assembly speaks as one Voice, the many speaking as one. For the first time in history, you are able to hear the actual Voice of Revelation speaking through God's Messenger. It is this Voice that has spoken to all God's Messengers in the past. The Word and the Sound of the Voice of Revelation are in the world and are available for you to hear in their original audio form.

THE MESSENGER is the one chosen, prepared and sent into the world by the Angelic Assembly to receive the New Message. The Messenger for this time is Marshall Vian Summers. Marshall is a humble man with no position in the world who has undergone a long and difficult preparation to receive God's New Revelation and bring it to the world. He is charged with the great burden, blessing and responsibility of presenting this Revelation to a divided and conflicted world. He is the first of God's Messengers to reveal the reality of a Greater Community of intelligent life in the universe. The

Messenger has been engaged in this process of Revelation since the year 1982.

THE PRESENCE refers to different but interconnected realities: the presence of Knowledge within the individual, the Presence of the Angelic Assembly that oversees the world or ultimately the Presence of God in the universe. The Presence of these three realities offers a life-changing experience of grace and relationship. All three are connected to the larger process of growth and redemption for us, for the world and for the universe at large. Together they represent the mystery and purpose of our lives, which the New Message reveals to us in the clearest possible terms. The New Revelation offers a modern pathway for experiencing the power of the Presence in your life.

STEPS TO KNOWLEDGE is an ancient book of spiritual practice now being given by God to the world for the first time. Steps provides the lessons and practices necessary for learning and living the New Message. In beginning the Steps, you embark on a journey of discovering Knowledge, the mysterious source of your inner power and authority, and with it the essential relationships you are destined to find. Its 365 daily "steps," or practices, lead you to a personal revelation about your life and destiny. In taking this greater journey, you can discover the power of Knowledge and your experience of profound inner knowing, which lead you to your higher purpose and calling in life.

THE GREATER COMMUNITY is the larger universe of intelligent life in which our world has always existed. This Greater Community encompasses all worlds in the universe where sentient life exists, in all states of evolution and development. The New Message reveals that humanity is in an early and adolescent phase of its development and that the time has now come for humanity's emergence into the Greater Community. It is here, standing at the threshold of space, that

IMPORTANT TERMS

humanity discovers that it is not alone in the universe, or even within its own world.

THE GREATER COMMUNITY WAY OF KNOWLEDGE is a timeless tradition representing God's work in the universe to reclaim the separated in all worlds through the power of Knowledge that is inherent in all intelligent life. To understand what God is doing in our world, we must begin to understand what God is doing in the entire universe. For the first time in history, The Greater Community Way of Knowledge is being presented to the world through a New Message. The New Message opens the portal to this timeless work of God underway throughout the universe in which we live. We stand at the threshold of emerging into this Greater Community and must now have access to The Greater Community Way of Knowledge in order to understand our destiny as a race and successfully navigate the challenges of interacting with life in the universe.

THE INTERVENTION is a dangerous form of contact underway by certain races from the Greater Community who are here to take advantage of a weak and divided humanity. This is occurring at a time when the human family is entering a period of increasing breakdown and disorder, in the face of the Great Waves of change. The Intervention presents itself as a benign and redeeming force while in reality its ultimate goal is to undermine human freedom and self-determination and take control of the world and its resources. The New Message reveals that the Intervention seeks to secretly establish its influence here in the minds and hearts of people at a time of growing confusion, conflict and vulnerability. God is calling us, as the native peoples of this world, to oppose this Intervention, to alert and educate others and to put forth our own rules of engagement as an emerging race. Our response to the Intervention and the Greater Community at large is the one thing that can unite a fractured and divided human family at a time of the greatest need and consequence for our race.

THE GREAT WAVES OF CHANGE are a set of powerful environmental, economic and social forces now converging in the world. The Great Waves are the result of humanity's misuse and overuse of the world, its resources and its environment. The Great Waves have the power to drastically alter the face of the world— producing economic instability, runaway climate change, violent weather and the loss of arable land and freshwater, threatening to produce a world condition of great difficulty and human suffering. The Great Waves are not an end times or apocalyptic event, but instead a challenging period of transition to a new world reality. The New Message reveals what is coming for the world and provides a preparation to enable us to navigate a radically changing world. God is calling for human unity and cooperation born now out of sheer necessity for the preservation and protection of human civilization. Together with the Intervention, the Great Waves represents one of the two great threats facing humanity and a major reason why God has spoken again.

HIGHER PURPOSE refers to the specific contribution each person was sent into the world to make and the unique relationships that will enable the fulfillment of this purpose. Knowledge within the individual holds their higher purpose and destiny for them, which cannot be ascertained by the intellect alone. These must be discovered, followed and expressed in service to others to be fully realized. The world needs the demonstration of this higher purpose from many more people as never before.

SPIRITUAL FAMILY refers to the small working groups formed after the Separation to enable all individuals to work towards greater states of union and relationship, undertaking this over a long span of

IMPORTANT TERMS

time, culminating in their final return to God. Your Spiritual Family represents the relationships you have reclaimed through Knowledge during your long journey through Separation. Some members of your Spiritual Family are in the world and some are beyond the world. The Spiritual Families are a part of the mysterious Plan of God to free and reunite all those living in Separation.

ANCIENT HOME refers to the reality of life and the state of awareness and relationship you had before entering the world, and to which you will return after your life in the world. Your Ancient Home is a timeless state of connection and relationship with your Spiritual Family, The Assembly and God.

About Marshall Vian Summers

Marshall Vian Summers is the founder of The Society for the New Message and the primary teacher of Greater Community Spirituality. For over four decades, he has been receiving a series of revealed teachings to prepare humanity for the profound social, political, and environmental changes unfolding worldwide and for humanity's emergence into a larger universe of intelligent life.

Originally working as a special educator for the blind, Marshall devoted many years to creating and teaching programs on relationships, inner guidance, and spiritual direction. In 1982, at the age of 33, he experienced a life-changing spiritual initiation in the deserts of the American Southwest. During this time, he encountered an Angelic Presence that asked him to receive a series of messages for the world.

Over the ensuing forty years, Marshall has followed a mysterious path, receiving what he describes as revelatory and prophetic messages meant to awaken individuals to their greater purpose in service to the world and to reveal the greater reality of life in the universe of which our world is a part.

Today, he resides in the Rocky Mountains of the United States, where he continues to teach people worldwide how to recognize and apply their innate spiritual intelligence to meet life's growing challenges and to prepare for humanity's contact with intelligent life from beyond our world.

Read more about the life and story of the Messenger
Marshall Vian Summers:
www.newmessage.org/story

Read and hear the original revelation The Story of the Messenger:
www.newmessage.org/story-of-the-messenger

Hear and watch the world teachings of the Messenger:
www.newmessage.org/marshallviansummers
www.youtube.com/marshallviansummers

The Voice of Revelation

For the first time in history, you can hear the Voice of Revelation, such a Voice as spoke to the prophets and Messengers of the past and is now speaking again through a new Messenger who is in the world today.

The Voice of Revelation is not the voice of one individual, but that of the entire Angelic Assembly speaking together, all as one. Here God communicates beyond words to the Angelic Assembly, who then translate God's Message into human words and language that we can comprehend.

The revelations of this book were originally spoken in this manner by the Voice of Revelation through the Messenger Marshall Vian Summers. This process of Divine Revelation has occurred since 1982. The Revelation continues to this day.

Hear the original audio recordings of the Voice of Revelation, which is the Source of the text contained in this book and throughout the New Message:
www.newmessage.org/experience

Learn more about the Voice of Revelation, what it is and how it speaks through the Messenger:
www.newmessage.org/voiceofrevelation

About The Society for the New Message

Founded in 1992 by Marshall Vian Summers, The Society for the New Message is a 501(c)(3) non-profit organization supported by readers and students of the New Message.

The Society's mission is to provide an education and preparation for humanity's emergence into the Greater Community—the larger universe of intelligent life in which we have always lived—and to expand human awareness and intelligence to make this possible.

With this, The Society presents the pathway for learning and living the New Message and supports people in beginning to take the steps to Knowledge in their lives.

With the support of volunteers, translators and financial supporters around the world The Society is able to make the books and teachings of the New Message available to people in over 35 languages free of charge as well as provide numerous educational offerings free of charge.

If you are inspired by this book and would like to be a part of making this message available to the world we encourage you to learn more about how you can help The Society by visiting newmessage.org/support.

The Society for the New Message

Contact us:

P.O. Box 1724 Boulder, CO 80306-1724
(303) 938-8401 (800) 938-3891
011 303 938 84 01 (International)
society@newmessage.org
www.newmessage.org
www.marshallsummers.com

www.alliesofhumanity.org
www.newknowledgelibrary.org

Connect with us:

www.youtube.com/thenewmessagefromgod
www.facebook.com/newmessagefromgod
www.youtube.com/marshallviansummers
www.facebook.com/marshallsummers
www.twitter.com/godsnewmessage

Donate to support The Society and join a community of givers who are helping bring the New Message to the world:

www.newmessage.org/support

Books of the New Message

God Has Spoken Again

The One God

The New Messenger

The Greater Community

The Journey to a New Life

The Power of Knowledge

The New World

The Pure Religion

Preparing for the Greater Community

The Worldwide Community of the New Message from God

Preparing for the Great Waves of Change

Steps to Knowledge

Greater Community Spirituality

The Great Waves of Change

Life in the Universe

Secrets of Heaven

Wisdom from the Greater Community: Books One and Two

Relationships and Higher Purpose

Living The Way of Knowledge

www.ingramcontent.com/pod-product-compliance
Lightning Source LLC
Chambersburg PA
CBHW020352170426
43200CB00005B/138